LEYTON SENIOR

BORROWER'S NAME		
Miss J. Martin (Jacinta)	5.S2.	17/1/72
Anne Burwood.		

PS.1236

National Association for Maternal and Child Welfare

YOUNG STUDENTS' BOOK OF CHILD CARE

THIRD EDITION

BY

LEONORA PITCAIRN

Norland Nurse, A.R.S.H.

CAMBRIDGE
AT THE UNIVERSITY PRESS

1973

Published by the Syndics of the Cambridge University Press
Bentley House, 200 Euston Road, London NW1 2DB
American Branch: 32 East 57th Street, New York, N.Y. 10022

© Cambridge University Press 1963, 1973

ISBN: 0 521 20122 5

First edition (by Mildred Lister) published
by Eyre & Spottiswoode 1948
Second edition (by Leonora Pitcairn) published
by the Cambridge University Press 1963
Reprinted 1967
Third Edition 1973

Printed in Great Britain by C. Tinling & Co. Ltd, London and Prescot

CONTENTS

Introduction *page* ix

1 Before the baby is born 1

   The expectant mother. Mental attitude. Good advice and ante-natal care. Teeth. Loose clothing and sensible shoes. Personal cleanliness. Relaxation, rest and sleep at regular hours. Fresh air and exercise. Social life. Food. Drink. The little troubles of pregnancy. Preparations. What a baby will need. Preparations should be made early.

   SUPPLEMENT. *Foundations of good health now. How a baby grows in embryo stage. Birth.*

2 The newborn baby 20

   First six weeks of life. Suggested timetable. Self-demand feeding. Choosing something between routine and self-demand. To help you understand the small baby. Bathing baby. Weight.

   SUPPLEMENT. *Some common worries. The navel. The fontanelle. Hiccups. The fretful infant. Spoiling. Bringing up a little milk after feeds, or vomiting. Constipation. Diarrhoea*

3 The child from three to six months 36

   How to feed during the first few weeks. When the mother is up and about again. How to overcome any difficulties there may be. Insufficient breast-milk. Things to remember in complementary feeding. Additional foods for the breast-fed baby. Artificial- or bottle-feeding. Manner of giving bottle-feeds. Choice of bottle to be used. Teats. Cleansing bottles. Cleansing teats. Sterilizing of bottles and teats.

Making up a feed using fresh cow's milk. Making up a feed using dried milk. Making up the day's supply at one time. Temperature of feed. Additions to the artificially fed baby's diet. Changes in baby's day from about three to six months. Companionship. Stopping the late feed. Further additions to his diet.

SUPPLEMENT. *Calculating quantities of food babies need. How to tell how much breast-milk baby is getting. How to decide the quantity he needs. Quantities for a bottle-fed baby. Dried milk. Over-feeding. Signs of under-feeding.*

4 The child from six months to one year old   54

Play-pen. Weaning. How to begin making additions to his milk diet. Some of the foods suitable for a baby during the weaning period.
Cleanliness and care of food is still important. When to stop giving sugar in his milk. Travelling.

SUPPLEMENT. *Signs of readiness for weaning. Companionship rather than too much direct attention. Spoiling. Clinging to mummy. Crying.*

5 Children up to school age   70

During the second year of life. Temper tantrums. Suggested day for a toddler. Food. Rest. Helping him to become capable and independent. Time to learn to be sociable, to mix with others. Getting him ready for school. Animal pets.

SUPPLEMENT. *More about toddlers. Examples of the different impressions a mother can make. When to offer choice and explanations. Curiosity. Imagination. Father's part.*

6 Safety and first aid   84

In the home. Preventing burns or scalds. Preventing dangerous falls. Preventing cutting or piercing. Preventing swallowing objects, or poisons. Preventing suffocation. Preventing accidents from electricity or gas. Safe area of play. Out of doors. In the garden. In the street. In cars. On holiday. First aid. For burns and scalds. Falls. Head injuries. For cuts, stabs, abrasions, wherever there is an open wound. When objects have been swallowed. Poisons. Lesser troubles which may need first aid.

SUPPLEMENT. *Accidents. More about poisons. Fractures.*

7 Diet and health   99

Food values. Foods for growth, or body building. Foods for protection and maintaining health. Foods for energy and warmth. Teeth. Cleaning of teeth. Visits to dentists. When the teeth come. Eyes. A squint. Injuries or foreign bodies in the eye. Feet. Posture. Clothing. Winter clothing. Summer clothing. Night-clothes. Toilet training. Bed wetting.

SUPPLEMENT. *More about toilet training. More about bed wetting. Vitamins. Calories. Digestive system.*

8 Sleep and general health   116

Fresh air. Sun. Appetite. Unwillingness to take milk. Crying. Crying for his mother.
SUPPLEMENT. *Emotional problems. Aggressiveness. Jealousy.*

9 Character training   134

The home. The mother. The father. Good habits. Bad habits. Truthfulness and lying. Why should a child lie? Obedience and discipline. Punishment. Good temper and self-control. Good manners. Fairness, unselfishness and thoughtfulness. What is meant by bringing up a child to be a good Christian. Religion.

SUPPLEMENT. *Object of character training. Appreciation of the child's view.*

10 Childish illnesses   150

   Nursing a sick child. Washing him in bed. Keeping him happy. Prevention. Hospital. Immunity. The common cold. Infectious diseases. Chicken pox. Mumps. Measles. German measles. Scarlet fever. Whooping cough. What to do. The premature baby. Twins. Adopted children.

   SUPPLEMENT. *Diseases. Smallpox. Diphtheria. Poliomyelitis. Giving medicine. Ear infection. Croup. Convulsions.*

11 Play   168

   Toys. Points in general about toys for small children. Baby, from two or three months old up to nine or ten months old. Respect for toys. Parents and play. Telling stories. Guiding a child's reading. Television. Music. Importance of duties. After about the age of six. How to feel at ease with children. Baby talk. Fantasy. Older children.

12 Speech and childish difficulties   185

   Speech. Baby talk. Stuttering and stammering. Nail-biting. Fears. The fatherless child. The working mother. The handicapped child. How to give him the desired attitude. Answering questions.

Introduction

It is accepted nowadays that more than instinct is needed in the care and upbringing of children. When a baby is born his parents do not find that they know automatically what they should do, and it seems right for the study of child care to be a part of general education. A basic understanding, and interest roused early, should pave the way towards success in this as in any other subject.

By the laws of nature, women have been chiefly responsible for dealing with young families. Today we believe also that the ideal is for the family to work as a unit, for the father to accept responsibility along with the mother. He is not only the breadwinner (and possibly the disciplinarian), but he should also take an intelligent interest, and help in whatever ways are possible. To do this he needs to be knowledgeable and not only to trust in his common sense and luck.

Obviously all who become parents should benefit from the study of children. Moreover, most women have direct contacts at some time or other with children, and men too may find themselves glad not to be entirely ignorant. There may be small friends who need attention in an emergency, nephews, nieces, brothers, sisters, or adopted children.

Girls who become nurses or teachers will find it an advantage to have had a good grounding for the greater skill needed in these professions. Even librarians, shop-assistants, hairdressers, and many others, will need to deal with and understand children. Within the circle of family and friends there is scope for putting knowledge to good use. In any community a woman will fit in more easily if she is not ignorant about something so tremendously important.

There are many reasons for studying child care which students will discover for themselves throughout their lives. There are the good, unselfish reasons which include thought for others, the wish to be helpful, useful, kind, understanding, efficient, and the wish to play a part in producing the best in future generations. As well as these there are the more selfish motives, which all the same are worth consideration. Would you like, when looking after children (your own if you have them, or anyone else's) to feel at a loss, to be worried and ignorant, to have your whole way of life turned upside-down? Babies and children can indeed be disturbing, but when properly handled should not be troublesome. No one need become harassed or overtired, if she (or he) knows how to manage in order to keep them well and contented. Without any knowledge, the coming of a baby into a household can create havoc! It can even bring misery, depression and discontent, instead of the joy and happiness that has been expected.

Study this subject then, and try to add throughout your

life to whatever you learn now. You are asked to consider the ideal methods of bringing up children. It may be that circumstances make certain things impracticable. Aim high all the same, when you are responsible for what is done. If you are helping parents who seem to you to be ignorant and careless, introduce your own opinions with tact.

Most important of all, remember that babies and children need *loving* care. The best food, most suitable clothing, efficient handling, cleanliness, fresh air, sunshine—all the things about which we hope to teach you—will be wasted if love is lacking.

Introduction to the Third Edition

In the third edition of this book much of the familiar content has been retained, but the opportunity has been taken to revise the book in the light of experience and changing patterns of child care. The publishers would like to thank Mrs Iris Beale for her help with photographs.

CHAPTER I

Before the Baby is Born

The expectant mother. Mental attitude. Good advice and antenatal care. Teeth. Loose clothing and sensible shoes. Personal cleanliness. Relaxation, rest and sleep at regular hours. Fresh air and exercise. Social life. Food. Drink. The little troubles of pregnancy. Preparations. What a baby will need. Preparations should be made early. SUPPLEMENT. *Foundations of good health now. How a baby grows in embryo stage. Birth.*

The first thing a baby needs is a good home. The size or grandeur is unimportant, but it *is* essential that a secure background of parental affection is provided. This can only be achieved where a marriage is based on the sort of mutual love which is of the enduring kind, not simply a fleeting passion.

We see, then, that those who dream of being parents need to decide on their marriage partners most carefully.

Good health, too, is something worth giving to our children. The healthy young people of today will be the healthy parents of tomorrow, so that the foundations of good health for future generations can be laid immediately by all young people.

Some are less healthy than they need to be because

they are careless or thoughtless about such matters as personal and household cleanliness, care of their teeth, opening their bedroom windows at night, fresh air and exercise, what they eat and drink, what clothes and shoes they wear, their forms of enjoyment. Regular daily attention to such simple affairs can soon become a habit and need not prove troublesome to carry out. It is not enough to be merely 'not ill'. We should all aim at feeling perfectly fit.

Many mothers say that the months during which their babies are coming are among the happiest of their lives. If you know such a mother, look carefully at her home, the way she lives and behaves, the sort of man she has taken as husband, the people she has chosen for friends and companions. It is likely that you will find that she and her husband have made a good home (not necessarily composed of the luxuries money can buy), that they live and behave properly, and that her family and friends are contributing in various ways to her enviable state of mind.

The expectant mother

Here are a few points to be remembered.

Mental attitude

The expectant mother should think in the right way about childbearing and childbirth. Many mothers-to-be naturally feel happy and confident, and look forward to the joy to come. Others may be nervous, even frightened, and in such cases every effort should be made to cultivate more sensible feelings. 'Old wives tales' should not be listened to—and certainly never believed. It is sometimes said that something which happens to the mother (e.g. seeing an accident) can produce certain effects on the baby. Such stories may be told in good faith but are the result of ignorance. In fact no shock or worry a mother has will affect the baby she is carrying, unless she herself is damaged in some way, or worries so that she becomes unwell. The baby, before birth, is nourished entirely through his mother's blood-stream, therefore the important point is for the mother to be healthy and to remain calm and contented.

Pregnancy is a normal and natural process. The expectant mother's condition is not one of delicacy or illness. She is more likely to feel well and happy now than at any other time of her life as long as she is sensible, takes suitable care of herself, and allows no stupid unnecessary fears to creep into her mind. This work which her body is doing now is the work for which it was intended. It is natural and right, and has a beauty all its own.

Childbearing, however, does impose a certain strain on the body and, for this reason, sensible attention must be paid to it at this time.

Good advice and ante-natal care

Advice should not be sought from all and sundry. As soon as a woman thinks she is pregnant she should visit her doctor or attend a clinic. In this way she will get the best advice possible, and if she carries it out there will be nothing for her to worry about. She should be ready to speak frankly to doctors, health visitors and midwives, for there is no need for embarrassment about such a common happening as having a baby.

Teeth

Even if it is not the usual time for a visit to the dentist, an appointment must be made at once, for teeth and gums are more liable to decay during pregnancy. Decayed teeth are harmful and should be prevented whenever possible and dealt with always in the early stages. Morning or evening brushing will not be enough. The ideal would be to clean the teeth after every meal, or at least to rinse out the mouth if brushing is not possible. Absolutely thorough and efficient rinsing of the mouth is important.

Loose clothing and sensible shoes

These will contribute much to the mother's comfort. Clothing should hang loosely from the shoulder and allow freedom of movement without constriction at the waist. Many mothers do not need a corset if the tummy muscles are strong. If one is used, a simple maternity corset with patent adjustments is desirable. It must be of a good make so that there is no restriction or discomfort of any kind. If a corset is not required, a light suspender-belt is needed, as garters may interfere with the circulation. Brassières must be worn, and it is important to get a larger size than usual if necessary, as tightness here would not be sensible.

Shoes should be well fitting and have low heels for safe and comfortable walking.

Personal cleanliness

Every effort should be made to keep the hair nice and the whole person fresh and clean. Underwear needs frequent washing. Clothes which cannot be washed may at least be hung out in the fresh air at times. If the general appearance is kept up to a high standard the mother will feel better, and be better than if slackness and untidiness is allowed.

Relaxation, rest and sleep at regular hours

This must become a routine. If the expectant mother does not relax easily she must try to find out how to do so, from her clinic or doctor. About eight hours each night of deep refreshing sleep should be the aim, and longer if need is felt for extra sleep. An afternoon rest, especially during the later months of pregnancy, is advisable. Tiredness should be avoided by careful planning of the day and sitting or lying down according to individual needs at certain times.

Fresh air and exercise

These are as important as rest, for laziness is one of the worst enemies of good health. Bedroom windows should be open at night and the house kept well ventilated. At least an hour a day (preferably more) should be spent out of doors. Walking is excellent at this time. As regards other forms of exercise such as cycling, swimming, tennis, games—it would be best to discuss these with the doctor. Generally speaking a healthy woman can lead her normal life and be as energetic as usual.

Social life

Sensible plans for social activities are good but too many late nights, or being in crowded or stuffy places, should be avoided.

Food

There used to be rubbish talked about the expectant mother having to eat for two. Nowadays most people know that this does not mean eating twice the quantity of everything, but of eating enough of the right foods for baby's growth. The day's diet should be planned so that each main meal has the right balance. You will learn what this means in a later chapter. The growing baby needs the right materials and these will be supplied if the mother includes the following in her diet:

a pint of milk in some form each day
plenty of fresh fruit, salads and vegetables
3 or 4 eggs a week, cheese
fish twice a week
liver once a week, red meats
bread, cakes, biscuits, puddings, sugar, cereals in moderate
   amounts
remembering of course foods containing vitamins.

It is best to stick to three regular meals a day and to try to resist craving for odd food at irregular times. Meals should be cooked properly and served attractively.

Drink

Plenty of water should be drunk, first thing in the morning and at any time during the day. This will help to keep the kidneys in good order. It will also be good training for when baby is being breast-fed, and an ample supply of fluid is advisable.

The little troubles of pregnancy

It is only right to mention that there is a possibility of certain minor discomforts. Though having a baby is not an illness, it is a time when the mother can benefit from a few extra attentions and consideration from those around her. She may feel responsibilities weighing heavily upon her, for to bring a new life into the world is no light matter. But if she is following the good advice we hope she has obtained, then troubles should not be too great.

Morning sickness is common and can often be helped by having a cup of warm sweet tea and a dry biscuit before getting up, and by avoiding foods which might cause squeamishness (e.g. large, heavy or fried meals last thing at night). Severe sickness can be helped by the doctor.

Indigestion and heartburn often occur. Again this may simply be something in the diet which does not suit the mother, and thought should be given to what might be causing it. Constipation must be avoided, if possible by eating such foods as prunes, fruit, greens, salads, wholemeal bread. If it becomes bad the doctor should be consulted about a suitable laxative.

All troubles may be taken to the doctor or clinic in their early stages. In this way the parents will be free from worry as they will be dealt with by experts.

Preparations

What a baby will need

There are no hard-and-fast rules but here are a few suggestions. Whatever is planned will, in any case, depend on the personal tastes of the mother and the amount of money she has to spend. If the birth takes place in a hospital, then that particular hospital should be consulted and all instructions followed. If the birth is to be at home then the midwife will say what she needs.

Bath and toilet requirements. As soon as he is born, baby will have to be bathed, and he will need to have a bath every day. There are many different types of bath—papiermâché, enamel, rubber, plastic, canvas; but a large basin (or even the kitchen sink) would do. Something which is deep enough to allow baby to splash is best and it should be on a firm stand or else on a strong table of the right height. Whoever is bathing him will need:

a nursing-chair
2 water-jugs (one for hot and one for cold water)
a bathing-apron
large bath-towel
small soft face towel
bucket for soiled napkins
baby soap
soft flannel
hairbrush
tray or basket with cotton-wool
small bowl for attending to eyes, nose and face

These are the essential things, but some mothers like to have more; for instance:

2 buckets, one for dirty another for wet napkins
a flannel for the face and another for the body
a basket to put soiled clothing in
an extra dish to put the pieces of cotton-wool in after use
powder
oil
ointment
bath-mat
little clothes-horse for hanging baby's clean clothes on
screen in case of draughts
baby's pot
bath thermometer
trolley to keep all baby's things on
also a tray for feeding-times

The clothes he needs. Before shopping for the new baby it would be as well to keep certain important points in mind. The *materials* should be:

warm
soft
light in weight
porous
non-irritating to delicate skins
good wearing
attractive in colour and design
non-inflammable

The *garments* themselves should be:

well cut and loose enough to allow free movement
free from constricting bands
simple and attractive in style
easy to put on and take off
easy to wash, dry and iron

Remember too, that though babies need to be kept warm in cold weather, it is very bad for them to be too hot. It is not likely that they will ever need more than three layers of clothing even in winter provided that the materials are carefully chosen. They should be dressed according to the weather rather than the time of year.

Here is a list giving the garments suggested. The numbers of each bought must depend on the money available for spending.

- 3 or 4 vests (wool, silk and wool, or cotton or stockinette for hot weather. The wrap-over kind are easy to manage but the important thing about style is that they must not be difficult to pull over the head).
- 3–4 stretch suits—babygrowers
- 2 dozen Turkish napkins
- 2 dozen muslin napkins (or 2 dozen Turkish and muslin combined napkins) and pins or disposable napkins or liners
- 3 or 4 cardigans
- 1 or 2 hats or bonnets
- 4 or 6 little bibs of towelling
- 1 large shawl and possibly 1 smaller one or 2 cellular baby blankets

All these garments may be made or bought in the second size, as baby will rapidly grow out of the first sizes and it will not matter if they are all a little big at first. Other things which will not be needed immediately, but which some mothers may like, are:

frocks
pram suits
knickers
leggings
gloves
handkerchiefs or soft paper tissues

Baby's bed and bedding. Some mothers love to get a really beautiful cot ready for their babies, lined and decorated in lovely materials and colours. Anything may be chosen provided care is taken with the points which really matter to baby himself. He needs to have a soft but firm and comfortable mattress to lie on, and sides to prevent him from rolling out. There are wicker cradles, carricots, and Moses baskets, of all kinds. Any of those may be used. The more easily it is carried about, the more convenient it will be. The most economical buy would be a full-sized cot with let-down sides. This would be all he needs till he is ready for a bed.

Interior-sprung reversible waterproof cot mattresses with fitted sheets are best. Whatever blankets are used they must be warm but light. Two or three cellular blankets are recommended. Eiderdowns are not necessary, nor are pillows. If a pillow is used it must be flat and firm to prevent danger of suffocation.

Perambulators. Babies spend much of their time sleeping out of doors so that we must consider this when buying a pram, and not simply buy what we might call a baby-carriage, for taking him about from place to place. It should therefore, be large enough for him to stretch out in comfortably until he is a few years old. Here are the points to note when choosing a pram:

Size right for baby, and not so deep that he does not get air.
Height of handle right for mother. It is so tiring to have to push a pram which is too high or too

low. The handle should come to about waist level, and a retractable handle is useful if storage space is cramped.

GOOD springs and tyres, absolutely fool-proof brakes and safety chain, strong fittings for safety straps.

Hood lining is best in a dark colour which is restful to baby's eyes.

Hygienic, easy to clean.

Pram bedding is similar to cot bedding and the same may be used for cot and pram though it will be easier to keep things aired properly if separate bedding is kept for night and day use. A sun-canopy or sunshade may be needed in the summer and a pram-net to protect baby from cats is desirable if he is to be left in the garden unattended.

Baby's room. We cannot all live in houses possessing an ideal room which can be spared for baby, but it might help to consider what the ideal would be if we could manage it. The very best arrangements which the home will allow can then be made.

The ideal situation is on the sunny side of the house with windows on two of the walls, preferably facing S.E. or S.W. so that baby will get the sun when there is any. If he is only to use this room at night, for sleeping in, this is not so important of course, and it is the room where he will spend most of his time which matters most. Washable walls are best, in a pretty, light shade and he will soon appreciate picture friezes. Cork linoleum or any floor covering which is easily washed, is advisable, but it should not have a cold surface as he will sit about on it often, as he grows older.

Curtains should be washable, bright, and gay, and there should only be the furniture which is really needed, leaving as much space for playing and romping as possible. Ample cupboards for toys as well as drawers for his clothes, a high-chair or one which hooks over any ordinary chair will be wanted as time goes on. A play-pen will be most helpful and useful and if this is used early baby will be more likely to be happy in it when he reaches the toddling stage and has to be kept in a safe place at times.

Heating arrangements for winter must be adequate and safe and yet baby must never be kept in a room which is too hot and there must always be plenty of fresh air.

Preparations should be made early

Babies often decide to come into the world before they are expected. Everything required for the mother and new baby should have been bought or made, then washed and put away carefully so that all is available for use at a moment's notice.

Household affairs should be planned too, so that when the moment comes the mother can relax contentedly and let those who have promised to help her, carry on until she is ready to take up the reins again.

Supplement

Foundations of good health now

You should now have a simple picture of the preparation for the birth of a child. Here are a few further points to consider if you wish to begin *now*, to prepare for the part you can play in the future.

Realize fully that what you do throughout your life will affect not only yourself but others in the future. You have foundations to lay, which cannot be done carelessly. Is it not rather an exciting thought that so much can depend on you? Can you really accept a popular idea I have sometimes heard, that your life is yours and yours alone, and that what you do with it is your own affair?

The basic attitude to life needs to be taken seriously. Use your own mind and education to plan for yourself what sort of person you want to be. Create the ideal in your imagination and determine to strive towards it.

The value of considering personal hygiene and general rules for good health comes into this. Good health is a step towards happiness; poor health a drawback. To a great extent this is in your hands. Throughout this book you will learn how to bring up children to be healthy. Apply it to yourself, remembering that these are not things we force on small children during the period that we are stronger than they are and can be said to make them do what we want. Rather is it our intention and hope to form good habits, which they will retain to their own great advantage. If you have been brought up with these habits you must realize that they are not something you can discard as soon as you are free from parental control. There is a reason for them. If, on the other hand, you feel that your standards are at present lower than those we show you, then determine to raise them as and when you can. It is never too late to do this. Apply what you learn, to yourself, to your daily life, if you would like to have the best chance of good health.

Hygiene, like child care generally, is not instinctively acquired. It has to be studied. For the good health of the individual the following are necessary (and these will be discussed throughout):

1. Healthy and proper mental occupation and outlook. This does not mean there should be no fun in life. But there should not be excesses either in work or play. Moderation in all things. (Have spare-time occupations and pleasures which satisfy and fulfil.)
2. Cleanliness in person and surroundings.
3. Fresh air, exercise, sun.
4. Well-balanced meals.
5. Work, relaxation, sleep and rest. A good general guide for an adult who wants to keep fit would be to divide the 24-hour day into three. Approximately 8 hours of work, 8 hours of relaxation and/or rest, 8 hours sleep. The amount of sleep and rest needed varies with the

individual. Most young people seem to need about 8 hours.

How a baby grows in embryo stage

A baby growing inside a healthy mother, whose father was fit and well, will have the best chance. This is something we owe to any life we dare to bring into this world. The development of a baby inside his mother is one of the most marvellous happenings.

During this period the *embryo* as it is called, passes naturally through the most remarkable stages. These are still not completely understood, although scientific investigations reveal more and more about the mystery of life before birth and how it can be safeguarded.

The beginning is the union of two tiny cells, from the male and female, and the baby starts to develop. He floats in a small sac filled with fluid which protects him. It gives him somewhere to move about, cushions him from the movements his mother must make in her daily life and guards him from cold and injury.

At first he does not in any way resemble a baby but by the time he is about seven weeks old he will have a head (large in proportion to his body at this time) and eyes, and his limbs will be beginning to form.

Nourishment is transferred from the mother's blood-stream, to the embryo, by the *placenta*, which also takes waste products from the baby back into the mother's blood-stream. These travel along the *umbilical cord* to the baby's navel and blood-stream.

By the end of the third month baby's bones are forming and he draws calcium from his mother. He will now be recognizable as a child and will begin to kick freely in his fluid-filled sac. The heart and blood-vessels will have developed and be filled with blood and for this he will have absorbed iron from his mother.

During the fourth and fifth month he grows in size and his teeth start to form. He is drawing further upon his mother for calcium and it is to be hoped for his sake that she is drinking plenty of milk.

By eight months he will be well developed, possibly almost plump, but he is not yet ready for birth though babies prematurely born at this stage have a good chance of survival.

By nine months he will have stronger sucking instincts, and not quite such a low resistance to infection. He will be more capable of standing up to changes of temperature, and better able to adjust himself to this strange new world.

Birth

When the time comes for the baby to be born the mother works to help him, hence this is commonly called being 'in labour'.

The way she works is this. At first her muscles contract and retract, and the opening of the womb becomes larger so that the baby's head can begin to come through. Then the muscles contract more powerfully, and the mother begins to push spontaneously or 'bear down' as they call it, in order to help her child to come out. Between these times of working along with the contractions of her muscles she is usually taught to relax.

When the child is born, the cord which has attached him to his mother is cut and he is now in a new world.

The mother's womb takes a short rest and then the placenta (sometimes called afterbirth), having finished its work, is expelled.

It will be about three months before the womb returns to its normal size and if the mother breast-feeds her baby she will be assisting this process.

CHAPTER 2

The newborn baby

First six weeks of life. Suggested timetable. Self-demand feeding. Choosing something between routine and self-demand. To help you understand the small baby. Bathing baby. Weight. SUPPLEMENT. *Some common worries. The navel. The fontanelle. Hiccups. The fretful infant. Spoiling. Bringing up a little milk after feeds, or vomiting. Constipation. Diarrhoea.*

Parents often have feelings of awe about birth and may even be a little afraid. Fear can produce or increase pain but attendance at relaxation and discussion classes will help to overcome unnecessary worries. The mother must remember that a birth can be the most natural experience, and a wonderful event.

While teaching child care, I have been offered tales of unhappy confinements. Many girls view the prospect with horror. I have even been told that I haven't seen what they have seen, or heard what they have heard! That may be so, but the fact remains that a birth need not be a dreaded affair. No one would deny that there have been unfortunate or miserable births. If you know of such happenings, try in the light of your own better education, to help others and to determine that if you are ever concerned, you will set about the job in the right way. This does not mean that you will be scornful towards those who have not had your advantages.

Mothers of today and tomorrow are indeed fortunate because of the great strides which have been made in maternity work. There are doctors, midwives, health visitors, clinics, dentists, Home Help services, Medical Officers of Health, local authorities, welfare foods, grants. Information about all of these is easily obtainable. An expectant mother need only seek help and guidance from those who are available and ready to give it. Then she must make full and reasonable use of everything offered, carrying out instructions and attending the clinic regularly.

Skill, knowledge and experience bring first-class treatment to all who ask for it.

There is no need for you, as students, to consider this too deeply at the moment. Simply remember the sensible attitude your studies have taught you, use your influence with anyone you think may need it, and remember one other thing. In the same way as you can now look back on certain childish fears you may have had when you were small, so you will develop still more fully and as you grow up, you will be able to face and accept the things that are natural to grown-ups.

First six weeks of life

Immediately after baby is born the mother will have a short time of rest. This will be a chance for her to make final plans in her own mind about how she is to manage her household in the near future. Within the first week or two baby will indicate how he means to behave, whether he is inclined to be troublesome or whether he is placid and easily managed. Also his size and weight can affect the plans made for him. So only now, after his birth, can his mother really sketch out some sort of trial timetable not forgetting that his day must fit in with the rest of the household. Only in a happy atmosphere will baby thrive as he ought, therefore he cannot be the only person to be considered. His father's wishes and opinions should be consulted.

Suggested timetable

About 6 a.m. He should be made clean, dry and comfortable, then put to the right breast. When this is empty he will need a short rest and should be held over the shoulder or in a sitting position, to allow him to get his wind up, before being put to the left breast. When he is satisfied, has got rid of all his wind, is comfortable and happy then he should be put back to bed again (or out in his pram in warm weather).

He ought to sleep or lie contentedly until *about 9.30 a.m.* Then he can have his bath (or wash if he is to bath in the evening) and have his clean clothes on for the day and he will be ready for his next feed, *about 10 a.m.* This time he should be put to the left breast first and the same procedure followed as before. By about 10.30 to 11 a.m. he should be comfortably settled in his pram in the garden and should be contented till almost time for his next feed which would be *about 2 p.m.*

Thus the mother has time for her other duties. Baby should be in the open air except in foggy weather or extreme cold, but near enough to be heard if he should cry. If he does cry then his mother can see if he has more wind or is wet. As soon as she has made him comfortable she should return him to his pram.

Before the 2 p.m. feed he must be made clean and comfortable again (and mother should remember to give him alternate breasts first). By about 2.30 to 2.45 p.m. he should be back in his pram until about 5 p.m. probably, but as he grows older he will demand attention earlier. He should have a short mothering hour before his 5.30 p.m. bath (or wash if bathed in the morning). The length of this will be according to the baby's wants (no doubt clearly expressed) and the amount of time the mother can spare.

After the 6 p.m. feed (approx.) he must be allowed to get his wind up as usual and settled comfortably in his cot and should sleep till his next feed, probably about 10 p.m. After this he should sleep soundly all night, though some babies need a night feed for the first four to six weeks of life. If he wakens at all he may be made comfortable as during the daytime but should then be firmly returned to his cot. The question of sound sleep will be discussed fully later.

A similar timetable will be suitable for the bottle-fed baby. These suggestions can be considered by the mother, as she rests after the baby's birth and she can try to see how to fit in her other duties. Thought and planning now will make things run more smoothly when she resumes her responsibilities again.

Mother's undivided attention should be given to the baby while he is being fed, whether by breast or bottle. This gives him the feeling of security he needs. He should always be held in a comfortable position.

Getting up wind. Some air is swallowed by babies as they feed, which collects in their stomach. This causes discomfort. Half-way through feeds it is best to give baby a short rest and to see if he can rid himself of any wind. A few moments will be enough and he may or may not get rid of wind then. But at the end of his feed he may need about ten minutes to get wind up before going back to his cot or pram. He can be held over the shoulder and his back patted, or may sit upright (well supported) and be swayed gently backwards and forwards to get rid of the wind, or have his stomach rubbed. If he does not seem to get wind up very easily it may help to put him in a lying position for a second and then lift him up again.

If he seems to find it hard to get wind up then there is no need to worry. Perhaps he has none. If he has, and cries later, he can be picked up then, and dealt with.

Self-Demand Feeding

Some people prefer to let the baby feed when he himself indicates that he wants it. So long as this is not carried to

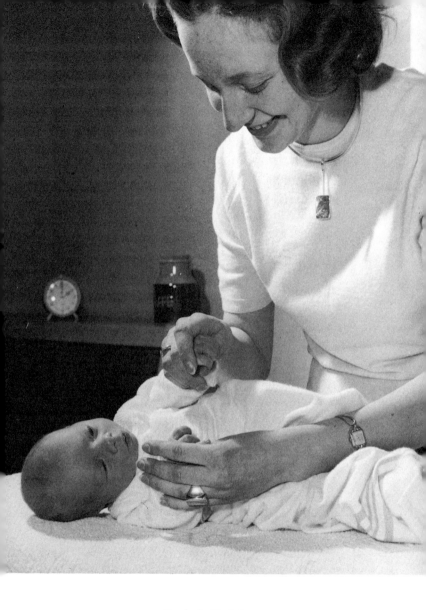

Ready for his feed

such extremes that the whole household is inconvenienced, the mother exhausted, and the baby thoroughly spoilt, it may turn out well. It may suit some parents better too.

Parents naturally want to do what is best for the infant but the parents themselves must also be taken into account. If the mother is tired by continuous feeding at frequent intervals, the father disturbed at night too, when he needs rest and sleep in order to do his job, then both of them will soon become low in health and spirits. This cannot be said to be in the best interests of the baby. So it would seem that some sort of timetable should be worked out, whether it is one made to suit the mother and the household primarily or one which the mother finds she can guide her baby into easily.

The infant who is being fed on self-demand may well settle down to more or less regular hours of feeding of his own accord. This does happen, but not always. If he continues to demand food at frequent irregular intervals after the first month it is likely that he will need to be helped into some sort of routine convenient to all.

There seems little doubt that self-demand feeding can be inconvenient, for the mother never knows when she will have free time for other duties.

Choosing something between routine and self-demand

The mother could work to a timetable which was flexible, to suit the baby, yet not so flexible that her whole life needed to be disorderly. As well as giving feeds earlier or later than planned according to what baby seemed to want, she could allow the baby an additional small feed rather than leave him to cry for lengthy periods, if she liked. This would not harm him though it might get him into the habit of asking for extra feeds, especially as he gets older.

To help you understand the small baby

The newborn baby does not move his body much even when awake. When he does move there is no purpose or control in his movements, yet he will cling to a finger placed in his hand and may draw his legs up when he cries.

His hearing is acute almost immediately. Loud noises will startle him. He will soon begin to know that different noises have different meanings. He will dislike some and be pleased by others. He soon begins to make use of his eyes as well as his hearing, to show interest and curiosity. It will not be long before he can distinguish objects and people.

When he cries he is practising the use of his vocal cords and preparing himself for speech later. When he moves in any way he is learning how to make use of his limbs, to understand about the space round him. His movements during his bath provide him with adequate exercise at this age.

He has been accustomed to the safety of his mother's womb, where he was held fairly tightly and comfortably. He will not like to be left to sprawl uncontrolledly till he has some control over his body.

Bathing baby

All preparations must be made before baby is picked up. Nothing is more upsetting than to find that something is missing, when baby is there, undressed, and ready. The bath and toilet requisites described previously must be collected and arranged.

Hands should be washed before touching baby's things or baby himself and a large bathing-apron worn.

In cold weather, windows should be closed and in a draughty house a screen used while baby is very small.

It is not possible to learn, from words, how to bath a baby, but it should not be difficult to arrange to watch or assist someone who is actually doing it.

When everything is placed conveniently ready, baby can be picked up and undressed. Rolled cosily in his large, soft, bath towel, his eyes, nose, ears should be inspected and cleaned as necessary, firmly but gently with small wisps of cotton-wool. Then his face washed with his soft flannel, and dried. Next with his head supported by the left hand, and held over the bath, the head is soaped, well rinsed, and then dried.

Usually baby dislikes this part of the proceedings, though if he is talked to and sympathized with he may not complain too loudly.

Next every inch of his body, every nook and cranny, should be washed with soapy hands, both back and front. Then the moment comes which he enjoys. He is ready to go into his bath. He should be supported by the left forearm under his head and shoulders and his left shoulder held firmly by the hand. The right hand supports his seat. Safely in the bath the right hand can be used to remove all the soap from his body and to splash him, which he will love.

After a few moments he must come out, though he will protest no doubt. The drying must be done carefully, with special attention paid to creases in neck, legs, groin, under arms. If he is well dried there will be little need for even a light dusting of powder, baby cream or ointment though these can be used where necessary; and he will be ready to be dressed.

Baby needs a bath every day, either in the morning or evening. If he is bathed in the morning he should be topped and tailed in the evening, or vice versa. Topping and tailing means that his eyes, ears, nose will be inspected and cleaned when necessary, his face washed, and his bottom washed.

At any time, if he is dirty his bottom should be washed, and before feeds it is necessary to make sure that his nose is clean for he will not be able to suck if he cannot breathe freely through his nose.

Everything to do with baby must be kept scrupulously clean.

Care of napkins. Immediately they are removed, wet napkins should be put into a pail which has been prepared for them by mixing an anti-bacterial and cleansing agent with water. Follow directions carefully.

Soiled napkins need to be brushed with a brush kept in

disinfectant for this purpose before being put to soak with the others. They should be rinsed thoroughly before being hung out to dry.

Weight

Babies lose weight in the first few days of their lives, until the mother's supply of milk is ready and the feeding established. This loss varies but is usually about $\frac{1}{2}$ lb. By the time he is one to two weeks old he should have regained his birth-weight and will then begin to gain.

It will be satisfactory for him to put on anything between 4 and 8 oz. each week (approx. 1 lb. a month) until he is about four months old and after that perhaps only 2–4 oz. each week. The exact weight he should gain cannot, of course, be laid down. To get an accurate weight-gain the weighing should be done at the same time each week, e.g. not before a feed one week and after a feed the next week.

Supplement

Some common worries

Most mothers, even experienced ones, know at some time or other, a horrid feeling of doubt or fear. Have they done everything they should? Is baby really all right?

New mothers tend to think of him as fragile and small and may find him almost terrifying.

The first thing is to recognize such feelings as natural and common to many of us. If the subject has been studied it is only a matter of gaining experience and confidence and becoming used to one's own child.

When a baby settles down and is what we call a good baby the mother is less likely to doubt herself but when he is a fretful infant she naturally tends to blame herself.

Mothers often worry unnecessarily about the following:

The navel

After birth, the umbilical cord was tied and cut close to the baby's body. The stump which was left withers and falls off in about a week or longer. It may leave a raw spot which may take a few days or even weeks to heal, but is of no importance. It only needs to be kept clean and dry. Unless the surrounding skin becomes inflamed, which is not likely, there is no need to seek advice or to worry, and the matter can be simply dealt with. Probably the doctor would recommend protection by dressings for a time and the use of antiseptic powder or powdered alum to help the process of healing.

The fontanelle

This is the soft spot on top of baby's head where the bones of the skull have not yet grown together. It is covered by a tough membrane which gives it some protection and it will not be damaged by ordinary handling.

Hiccups

Most babies hiccup quite a lot in the early months. There is nothing to do for this and it does not seem to be harmful. A drink of warm water may be given if the mother is worrying.

The fretful infant

It must be recognized that this is not an uncommon problem. Most babies are fretful at some time or other, for no apparent reason. They may sleep very soundly for long stretches, hardly waking even for feeds and then go through long fretful periods when the mother feels that they have never slept at all. It is difficult to give reasons for this. It could be that his nervous system has not yet settled down and he feels irritated for no reason we can put right. It could be that he has indigestion because of the immaturity of his digestive system.

Although this sort of fretfulness is nothing to worry about (provided he is not underfed, is progressing well and being cared for properly in every detail) the mother may not be able to stand it. For her sake it may be best to try to keep him quiet, by pushing him in his pram, taking him for a car ride even. A hot-water-bottle may comfort him, he may keep quiet if he has music to listen to.

Spoiling

This leads us to the question as to whether it is possible to spoil a baby in the first few weeks of his life. The experts do not think so though it is a different matter when he is two or three months old. So if a mother is worried by a fretful baby then perhaps it may be best to let her spoil him a little during the first few weeks. But if she can be persuaded that he can be left alone once she is satisfied that there is nothing seriously wrong and that she has done everything possible to make him happy, this will not harm him.

Bringing up a little milk after feeds, or vomiting

Again, most babies do this. Some bring up only a very little, very occasionally, others quite a lot after every feed. It may alarm the mother if baby brings up a lot of his feed but it is not serious in itself if he is otherwise healthy. The amount vomited looks larger than it really is and mothers often worry about whether he will be hungry and need more food immediately. Probably it is best to let his stomach have time to settle down. If he is gaining weight steadily then there is no need to worry if nature gets rid of some of his food.

Another point which often makes mothers anxious is when they notice that the milk he brings up is curdled. When it is understood how the digestion works this does not seem odd. For the stomach secretes acid to aid the digestion and the effect of acid on milk is to curdle it.

There are times, of course, when vomiting has to be

taken seriously. For instance if baby has been regularly and persistently sick after every feed since birth or if there is green bile in it, or if the food is vomited out with great force so that it lands at a distance from the baby's mouth, or if he has never vomited much before and then suddenly vomits a great deal. In such cases the doctor must be consulted.

Constipation

This is perhaps a difficult word to use about a small baby for they do not usually suffer much in this way especially if breast-fed. Mothers often think that they do because they do not necessarily perform every day. One baby may soil his napkins frequently, another only once in every three or four days.

In the breast-fed baby there may be very little waste matter, his food is so perfect for him, but if when he has a motion it is hard then he can be given extra fluid, 2 or 3 oz. a day. This should be boiled water either alone or with a small amount of sugar, perhaps half a teaspoonful, or it can be flavoured with a few drops of fruit juice.

In the bottle-fed baby the addition of a little extra sugar to his feeds may help and he too can be given extra fluid in the same way as the breast-fed infant. Extra sugar should not be given too freely, however, because of starting a habit which could later contribute to obesity.

Diarrhoea

This is more likely to occur where a baby is bottle-fed.

It can be caused by infection or from germs taken through the mouth and a strict check-up on cleanliness of his utensils and method of preparing his feeds will be needed.

It can also be caused through some other illness and will clear up when this has gone.

In any case, if severe, persistent or frequent a doctor must be consulted.

CHAPTER 3

The child from three to six months

How to feed during the first few weeks. When the mother is up and about again. How to overcome any difficulties there may be. Insufficient breast-milk. Things to remember in complementary feeding. Additional foods for the breast-fed baby. Artificial- or bottle-feeding. Manner of giving bottle-feeds. Choice of bottle to be used. Teats. Cleansing bottles. Cleansing teats. Sterilizing of bottles and teats. Making up a feed using fresh cow's milk. Making up a feed using dried milk. Making up the day's supply at one time. Temperature of feed. Additions to the artificially fed baby's diet. Changes in baby's day from about three to six months. Companionship. Stopping the late feed. Further additions to his diet.
SUPPLEMENT. *Calculating quantities of food babies need. How to tell how much breast-milk baby is getting. How to decide the quantity he needs. Quantities for a bottle-fed baby. Dried milk. Over-feeding. Signs of under-feeding.*

It can be said that most women can wholly, or at least partly, breast-feed their babies and this is what nature intends.

A mother should therefore assume that she can and will breast-feed, for a good mother wants to give her child every possible advantage in life.

Here are some reasons for breast-feeding which everyone should know:

1. It is the surest way of all of giving the infant the feeling of love and security he needs and forms a strong bond between mother and child.
2. Being provided by nature it is the only milk which is exactly right for baby. It contains all the foodstuffs baby needs for growth and development in the first few months of life.
3. It is pure, clean and fresh.
4. It is always exactly the right temperature.
5. It helps him to resist disease and there is no danger of infection.
6. Once established, it is easier than bottle-feeding and saves a great deal of time and trouble.
7. It costs nothing, though of course the mother's diet has to be good.
8. It is always ready when needed, and wherever the mother and baby happen to be.
9. It helps to form good teeth and well-developed jaws.
10. It is good for the mother.

Even if there are difficulties during the first few weeks, the mother should persevere. Doctors and midwives will always help and it is well worth a little trouble at first. Though most mothers have milk for baby when he is three or four days old, some take a little longer, but it does not mean that they will not be able to feed him at all.

The nursing mother should be taking all the necessary foodstuffs as she did during pregnancy. Plenty of fresh fruit and vegetables must be included, and a pint of milk a day unless she really cannot take it. Water must be drunk. It is good to get into the habit of having a large drink about half an hour before each feed or even at feed-time—milk, milky drink, tea, fruit drink or plain water.

After the first ten days or fortnight she should be going out for fresh air and exercise again, and her life should be a healthy one, with as little strain or worry as possible.

Most of all the mother should *want* to be successful in feeding her infant satisfactorily.

How to feed during the first few weeks

In bed the mother must be in what is a comfortable position to her. Often she can relax in a lying position on her side with baby placed alongside in exactly the right place for sucking. Or she may sit up, well supported by pillows. It is important that she should be completely relaxed. If she is not, she should move, even if it means interrupting baby for a moment, and get into the perfect position.

When the mother is up and about again

It is advisable for mother and baby to be alone quietly in a room without frequent noisy interruptions. The reason for this is, that to be entirely successful, both baby and mother should be able to concentrate on the job. A low chair will be convenient for the mother, and baby should be lying easily in her lap with his head in the crook of her arm. *Before breast-feeding the mother should wash her breasts.*

How to overcome any difficulties there may be

If all the advice sought and given has been taken, there are not likely to be serious difficulties. If there are:

1. The mother should go over again all the advice for successful motherhood and feeding, and make quite certain that she is neglecting nothing.
2. If she can find nothing wrong after these searchings, she should consider first the child and ask these questions: Has he any soreness in the mouth which might prevent him sucking? Could there be any malformation of his mouth or lip? Could his breathing be obstructed in any way?—thus making it impossible for him to suck and breathe at the same time. If so, will cleaning his nose before each feed help? Could he have any organic disease which makes him too weak to bear the strain of sucking?
3. If she still cannot find a satisfactory answer she can now consider herself: Is she holding the child properly so that his nose is not pressed against her, making it impossible for him to breathe? Has she any breast or nipple infection? Has she badly formed nipples which may make feeding difficult? If so, has she done all in her power to improve them?

Some of these questions might have to be answered by a doctor but the matter should be gone into fully.

There are, however, cases—extremely rare—where no reason can be found for difficulty. The mother appears to have an ample supply of milk and the child appears to be healthy in every way, but he simply does not and cannot suck properly. The instinct may seem to be lacking. In such a case much depends on the effort of the mother and the ability of the nurse handling the case.

Insufficient breast-milk

No mother should ever imagine that her supply of milk is not adequate without consulting doctor or nurse. If, however, it is proved that her supply cannot be immediately increased and baby is not thriving as he should, then the next best thing is to give complementary feeds. (The quality of breast-milk remains amazingly constant, though the quantity may be inadequate.)

Things to remember in complementary feeding

1. The breast-feed must always be given first and both breasts must be given. (Left side first at one feed, right side first at next, and so on, as usual.) The complementary feed must be given immediately afterwards. It should be prepared and all ready to be given as soon as the breasts are emptied and baby's wind got up.
2. Sucking at an empty breast must not be allowed, as this will give him wind, and upset him.
3. Even while complementary feeds are being given it

must be remembered that every effort can continue to be made to maintain and increase the supply of breast-milk. The aim should be to decrease the complementary feeds if the breast-milk can be increased and eventually they may be stopped. Remembering this it would be as well not to make the complementary feed too sweet or he may begin to refuse the breast, or to let him get the milk from his bottle too easily from a too large hole in the teat.

For the quantity to be given in complementary feeds it will be best to get expert advice if possible.

In giving complementary feeds all the advice in connection with bottle-feeding must be followed meticulously.

Additional foods for the breast-fed baby

Babies need extra vitamins and these are usually given in cod-liver-oil and orange juice which he may start taking at about a month to six weeks old.

Apart from these he will need nothing else until he is ready to commence weaning, but may have drinks of plain boiled water or sugar water. *All utensils, bottles, teats, and spoon if used, must be made and kept perfectly sterile.* (See instructions for artificial-feeding.)

Artificial- or bottle-feeding

We have shown that breast-feeding is best whenever possible but this does not mean that baby will not thrive if fed on a bottle. It only means that there are risks which a breast-fed baby does not have. These can be minimized if the greatest of care is taken in the selection, preparation and giving of his bottle-feed.

The usual substitute for human milk is cow's milk, either fresh, dried, or evaporated. It has to be adapted so that it is as near human milk as possible. It is best to be guided as regards consistency and quantity, by whoever has been helping with advice for baby, for this is really something for the experts to decide, in each *individual* case.

Manner of giving bottle-feeds

The bottle-fed baby has the same need of love and security as the breast-fed baby and *his feeds should be given to him in the same loving way*. (Keep the bottle tilted so that the teat is always full, to prevent air-swallowing.) To prop a bottle alongside him in pram or cot will be to deny him something he needs and can be dangerous.

Feeding bottle

The upright shape which can stand or lie flat is most convenient. A 9 oz. bottle, easy to clean, fill and hold is recommended—in lightweight polycarbonate which resists staining and is almost unbreakable. It should be complete with its own teat and screw top and cap.

Teats

A great variety of teats are on sale today and those as much resembling the human nipple as possible, will be best. A good test is to hold the bottle upside-down, when the milk should drip out steadily. Some infants need larger holes than others, depending on their sucking powers. If one sort of teat does not seem to suit baby then another should be tried.

Cleansing bottles

Immediately after the feed the bottle should be rinsed with cold water, washed in hot soapy water using a bottle-brush, then rinsed again in clean hot water. There should be no particles of milk, or film, left.

Cleansing teats

The teat should be turned inside out and rubbed with salt, then rinsed thoroughly in cold water.

Sterilizing of bottles and teats

The simplest and most usual way nowadays is for the bottles and teats to be immersed in a solution of hypochlorite. The maker's instructions should be followed strictly. When it is time for the bottle to be used it should be emptied of the solution and the feed poured directly into it. A fresh solution should be made daily.

Another method is for the bottles to be put into a pan and completely covered with cold water. After they have been brought to the boil they should be boiled for ten minutes. The teats should be put in for the last three minutes. Bottles and teats can lie submerged in this water, covered by a lid or muslin till they are needed. Or they can be removed from the pan, the teat firmly screwed inside the bottle, with the cap firmly in place. Or if bottles are not screw-topped then they must be kept covered and sterile and the teats stored in a sterile jar. *Whatever method is used it must be carried out correctly if it is to be effective.* Carelessness can be dangerous.

Utensils used for preparation of feeds must be clean and sterile too.

Personal hygiene of the person carrying out this duty is of great importance. Hands and nails should be immaculately clean, and indeed her whole person should.

Making up a feed using fresh cow's milk

A young baby would have difficulty in digesting whole cow's milk and it has to be modified by the addition of water. Since cow's milk has less sugar than human milk, sugar has to be added. The exact quantities and proportions for each individual infant will need to be worked out by the doctor or nurse.

The milk and water should be brought to the boil and the sugar added at the last moment. It should then be poured straight into the bottle which must be kept covered and free from germs till cool enough to give to baby.

Making up a feed using dried milk

Again the quantities and proportions for the individual child must be found out.

The required amount of powdered milk and sugar

should be put in a sterile jug, mixed to a smooth paste with cool, previously boiled water. Then hot boiled water added to make up the required quantity.

Making up the day's supply at one time

If it is not convenient to make each feed freshly just before feed-time, a requisite number of bottles must be bought. These should all be filled and covered as soon as the milk has been prepared. They must be kept in a cold place, preferably a refrigerator or ice-box. Then they need to stand in a pan or jug of hot water for about ten minutes before feed-time. In really hot summer weather it is advisable to make each feed freshly unless a refrigerator or ice-box can be used. When feeds have been stored the bottle must be well shaken before offering it to baby.

Temperature of feed

The feed, again to resemble human milk, should be about body-heat. The simplest way is to shake a few drops on to the inside of the wrist or back of the hand. If this feels too hot, cool it down a little in cold water, if too cold return the bottle to hot water again. Alternatively a chemical or electric bottle warmer may be bought.

Additions to the artificially fed baby's diet

Babies need extra vitamins (see p. 41). Cow's milk,

fresh or dried, only contains small quantities of these essentials.

Vitamins A and D are usually given in the form of cod-liver-oil, Vitamin C in orange juice.

Quantities for the individual child at the right age will be recommended by doctor or nurse and must not be forgotten.

Preparation. Again it is stressed that utensils must be *sterilized and kept sterile*, and everything which is given kept clean, pure and fresh.

Orange juice, whether fresh or bottled needs the addition of cooled boiled water. If bottled orange juice is used, care must be taken that it has not been pasteurized.

Boiled water. Whatever may be needed for the day only, should be boiled, so that it is prepared freshly each day.

It should be boiled for three minutes and kept in a sterile bottle for use as needed.

Changes in baby's day from about three to six months

The suggestions for his day vary little from those made when he was younger. General care should be as before. But as he begins to indicate his needs, gradual changes must be made, to suit his increasing intelligence as well as his growing body.

These will be, chiefly:

The giving of more companionship.

The beginning of toilet training.

The stopping of the late feed.

The addition of baby foods to his milk diet.

Companionship

The well-loved, contented, secure baby will not need to be cuddled and handled during all his wakeful hours, but he will love companionship. He can get this in various ways. He may be satisfied for most of the time to have toys for companions, many babies are. But as he sleeps less it may be best to bring him close to his mother for longer periods than before, e.g. half an hour or so before feeds especially if he is fretful. He can have his pram outside the window where she works and carry on some sort of conversation with her, or may be brought in to kick on the floor. It is quite a good idea to start putting him on the floor inside his play-pen so that he gets used to it before he reaches the age of objecting to anything new.

He will love the feeling of companionship as distinct from the 'mothering hour' we talked of before, which will need to be increased to a longer 'hour' as he gets older. That he should have a certain time of the day devoted entirely to him, when he is played with, cuddled, petted, spoilt a little, will become increasingly important.

Stopping the late feed (10 p.m. approx.)

Possibly baby himself will indicate that he no longer wants this feed but would prefer to sleep right through the night. If he does, it is better not to force it on him.

Or the mother may wish to discontinue it. In this case she can just leave him to sleep for as long as he will and give him only a small feed for a few nights at whatever time he does waken, perhaps midnight at first, and then gradually getting later. If he seems unwilling to give it up but the mother is determined that he should, she may try giving him a drink of water in the night to get him off to sleep again. He will usually be quite easily led into doing without it.

If he is breast-fed he and the mother will naturally adjust themselves so that he will get as much in the four feeds as he did in the five. If he is artificially fed the whole amount of food which he has been having in the twenty-four hours must be distributed in the four bottles, so that he is not getting less than before. He may refuse the extra in his bottles at first and should not be forced. He will soon adjust himself.

Further additions to his diet

Firstly, if he has difficulty in sleeping through the night when the late feed is stopped, he could be given a small quantity of baby cereal. Half to one teaspoon at first mixed with a little sweetened milk and water will be enough to give him, after his breast- or bottle-feed; or mixed with a flavouring of rosehip.

The additions of solids to his diet is the beginning of weaning and will be discussed fully in the next chapter, but this usually starts at some time between two and six months of age. It is best to be guided by the experts.

The advantages in starting sometime between two and six months are:

1. It can be done so very gradually as he will not *need* much at this early age.
2. He may take more easily to the new food and method of eating before he has strong ideas of his own.
3. He can get e.g. iron, which is scanty in milk.

Most babies at this age will be indignant if offered something in a spoon when they have expected to suck. So it is best to offer him a little of his milk first and then his solids, followed by the remainder of his milk.

Supplement

Calculating quantities of food babies need

Although we are fortunate in this country, in being able to obtain excellent advice simply by asking for it, it is of interest to understand how decisions are formed and conclusions reached.

In the matter of breast-feeding for example. This may go smoothly and easily and present no problems. Or the mother may have justifiable or even non-justifiable doubts. If she has they must be looked into.

How to tell how much breast-milk baby is getting

Test weighing. He should be put on the scales immediately before his feed (fully dressed). The weights should be left on the scales and he should be weighed again immediately after the feed (in exactly the same clothes of course and without any change of napkin). In this way the exact quantity of food he has taken may be measured.

This should be done, certainly over one period of twenty-four hours, and if possible over two, or three, twenty-four-hour periods, to get a fair estimate.

How to decide the quantity he needs

If he is putting on weight regularly and thriving such a calculation would be pointless, but it can be useful in certain cases.

First take the weight of the infant at birth and allowing for loss of weight in the first few days, work out what he ought to weigh approximately. It has been said that he should put on about 4-8 oz. per week. For the purpose of this calculation then, suppose we say that he should put on 6 oz. a week and work out what his weight might be.

Having done this, the amount of breast-milk he might require in twenty-four hours can be worked out as follows: For each pound of body weight allow $2\frac{1}{2}$ oz. of breast-milk, i.e. if baby should weigh 8 lb. he will require 20 oz. in twenty-four hours.

There is a tendency today to base calculations on the requirement being anything from $2\frac{1}{2}$ to 3 oz. to each pound of body weight, i.e. if baby should weigh 8 lb. he will require anything from 20 to 24 oz. in twenty-four hours.

He can if wished, be given the opportunity to take the larger amount but should not be forced.

It must be stressed that such calculations can only be a guide. It cannot be repeated too often, that no two babies are alike.

Quantities for a bottle-fed baby

The quantity of milk mixture which he is likely to need is worked out in the same way as for the breast-fed baby but the preparation of the mixture has to be studied too.

The following table shows the composition of cow's milk and human milk.

	Cow's milk %	*Human milk* %
Water	87	88
Protein	3.50	1.25
Fat	4	3.50
Milk sugar	4.50	7
Mineral salts	0.75	0.20

This shows clearly why it is necessary to modify cow's milk if baby is to be able to digest it and thrive on it.

One way of working out the mixture is to say that baby will need $1\frac{3}{4}$ to 2 oz. of cow's milk per pound of body weight and one level teaspoon of sugar per pound of body weight. This should then be made up to the required quantity of fluid, by the addition of water.

E.g. an 8-lb. baby would need:

14–16 oz. of cow's milk
8 level teaspoons sugar
6–8 oz. water

Thus his total quantity in twenty-four hours would be from 20 to 24 oz.

Dried milk

A measure (or a heaped teaspoon) of full-cream dried milk dissolved in 1 oz. water reconstitutes 1 oz. of cow's milk. Therefore the amount can be calculated in much the same way as for fresh cow's milk.

E.g. an 8-lb. baby would need:

14–16 measures of dried milk
8 level teaspoons sugar
water to bring up the quantity to 20–24 oz.

Over-feeding

This is seldom the cause of trouble and is rare in a breast-fed infant who usually ceases to suck when he has had enough, and gets rid of excess after a feed.
Advice will be needed however if there is:

Excessive gain in weight.
Vomiting after feeds—not that which is produced when baby brings up wind.
Frequent loose motions—showing curds and making the buttocks sore and red.

Signs of under-feeding

Persistent crying during the day and often at night. (Yet not all under-fed babies cry. Some become thin without as much as a whimper.)

Over a period of more than one week, loss of weight or only a very small gain.

Baby will be noticeably thin at bath-times.

Generally discontented.

CHAPTER 4

The child from six months to one year old

Play-pen. Weaning. How to begin making additions to his milk diet. Some of the foods suitable for a baby during the weaning period. Satisfactory diet for one year old. Cleanliness and care of food is still important. When to stop giving sugar in his milk. Travelling. SUPPLEMENT. *Signs of readiness for weaning. Companionship rather than too much direct attention. Spoiling. Clinging to mummy. Crying.*

By the time baby is about six months old his movements will be more controlled and will often have purpose. He will be able to touch and grasp whatever is within his reach. He will be able to sit up for short periods and any time now he may try to pull himself into a sitting position if he has not previously done so.

He will be responsive, smiling, laughing, chuckling or expressing displeasure. He will recognize people, show likes and dislikes and give many other signs of growing intelligence and comprehension.

After he is able to sit up alone he will soon try to crawl or get around in some way or other when he is put on the floor. By the time he is a year old, perhaps sooner, perhaps later, he will be pulling himself on to his feet and beginning to walk. The sounds he has been practising for so long may be recognizable, in small easy words such as 'Mum' and 'Dad'. He will understand much that is said to him and certainly the words 'yes' and 'no' will now have meaning.

Play-pen

At this stage a play-pen can be useful and is invaluable to the busy mother, but its usefulness and desirability must be fully understood.

It would be horrible to cage the baby continuously and to look on the pen as a place where baby can spend most of his time without anyone bothering about him.

But it is essential that he has somewhere to be perfectly safe and happy while mother is busy. If he has this, she will be able to hurry through her work and be ready to play with him or take him out all the sooner.

It is useful too as he can pull himself up and walk round holding on to the sides. He will be happy so long as he can experiment safely with all those new things he wants to do.

As he nears his first birthday he will be needing less and less sleep and will have to be more fully occupied during his waking hours. Suitable toys must be provided, more companionship, longer and more interesting walks. It is better not to force the pace of his development in any way or to strain his powers of understanding. He should

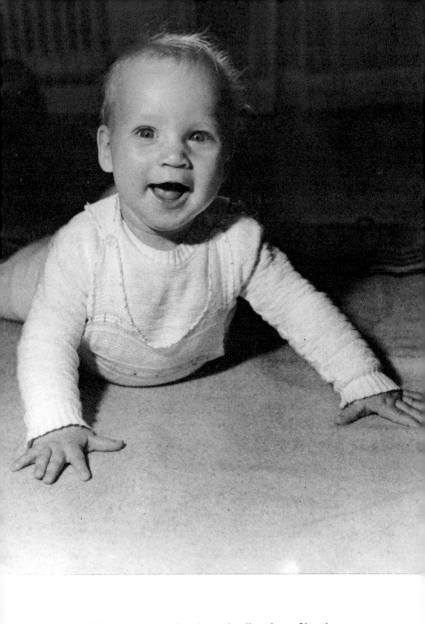

While mummy is watching he can be allowed out of his play-pen

grow naturally, at his own pace whatever this is, simply being given every opportunity.

Weaning

This means changing a baby's food from a milk diet which he has obtained by sucking, to a diet containing solid food, which he must bite and chew, and liquids which he drinks from a mug or cup.

This may well have been started before he is six months old by the giving of certain foods in small quantities along with the milk he was sucking. Nowadays it is quite common to introduce solids early, there is no definite age for this. Some people wait until he is 14 to 15 lb. in weight, others decide to start when he is only 11 to 12 lb. Many factors decide this, chiefly the individual baby's hunger and digestive system.

It is most usual however for baby to start on solids not younger than two months and not later than six months old.

How to begin making additions to his milk diet

Whatever age or weight he is, whatever kind of baby he shows himself to be, his introduction to new foods must be gradual. He should have a mere taste at first and quantities only increased as he seems able to enjoy and digest them.

There are several good ways of weaning a baby. Some people begin by giving sieved vegetables first, others make cereals the first addition to his milk diet. The following is simply an example of one well-tried method:

Baby is nearly five months old and weighs 15 lb.

6 a.m. or early morning feed. Breast or bottle as usual.
10 a.m. Half of the breast- or bottle-feed (approx.). A teaspoon of baby cereal, or less, mixed with a little boiled milk and sugar. Remainder of breast- or bottle-feed.
Other feeds as usual.

About three or four days later

Early morning feed as usual.
10 a.m. feed as above.
2 p.m. feed as usual.
6 p.m. feed with the addition of a teaspoon of cereal as at 10 a.m. (Different cereals used, for variety.)
10 p.m. feed may have been stopped by now. If not, give as usual.

About three or four days later increase the cereal given at 6 p.m. to as much as two teaspoons if he still seems to want his late feed and this will help him to sleep through the night without it.

When he is happily and comfortably taking the cereal at 10 a.m. and 6 p.m., an addition should be made to the 2 p.m. feed, of some of the following: bone or vegetable broth, meat gravy, egg-yolk, sieved carrots, spinach or cabbage, fruit purées, milk puddings.

Gradually he should be offered less of the milk-feed first and more of the solid food, so that the solid food eventually takes first place and the breast or bottle second place.

By the time he is about six to seven months old he would be having something like this:

Early feed as usual.
10 a.m. (approx.). Cereal, according to what he likes and can digest, followed by breast or bottle.
2 p.m. (approx.). Broth or gravy with sieved vegetables, small purée of fruit and milk pudding. Whatever quantity of milk he wants from breast or bottle.
6 p.m. (approx.). As at 10 a.m.

Now is a good time to wean him completely off the breast or bottle. He will probably have learnt, at a much earlier age, to drink from a mug and so it should not be too difficult to take him away from breast or bottle. In any case as his solid-food intake increases he will be likely to cease to want to suck.

The early morning feed can be replaced by a drink of fruit juice out of a mug and if he seems displeased about this he may like a rusk to suck, but he can only be given this when he is able to sit up, so that he will not choke himself.

The midday feed (about 2 p.m.) can be a substantial meal so that he will only need water to drink.

If he is breast-fed, the mother will now only be feeding him at 10 a.m. and 6 p.m. This gradual cutting out of breast-feeds will be best for her too.

The next stage will be to give him milk to drink out of his mug only, at 10 a.m., after his cereal, and the final stage to replace the breast or bottle with milk from his mug at 6 p.m. too.

When he is fully accustomed to these changes, digesting the quantities given him, and really on three meals a day which we can now call breakfast, dinner and tea, the times may be altered to suit the household, for the baby who eats from a plate and drinks from a mug is a different personality from the one who had to be fed from breast or bottle.

Some of the foods suitable for a baby during the weaning period

All baby cereals and baby foods, as much variety as possible.

Custards, blancmanges, semolina, junket, jellies.

Fruit purées of blackcurrant, apple, prune, ripe banana.

Egg-yolk, then lightly boiled or scrambled egg, meat gravy or broth, then minced or pounded fresh meat, liver, chicken, white fish, grated cheese.

Sieved vegetables, chiefly spinach, cabbage, carrot, then potatoes.

Rusks, plain or buttered or spread with honey, treacle or jam, then toast or thin sandwiches of brown bread and butter.

Sponge-fingers.

Some of the foods suitable for a baby during weaning period would be: fruit juice on waking, rusk or crust if he still wanted it.

Breakfast

Cereal with sugar and milk.

Egg two or three times a week or bacon and fried bread, or simply buttered toast and Marmite or marmalade.

½ pint milk, including what he had with his cereal.

Dinner

Meat or fish course with mashed vegetables (no need to continue sieving now).

Milk pudding of some sort most days, with stewed fruit or fresh fruit or fruit purée.

Light sponge-pudding with custard sometimes.

Tea (This should not be too early, say about 4.30–5 p.m. as it is his last meal.)

Thin bread and butter (mostly brown), with Marmite, grated cheese, honey, treacle, jam, lettuce, tomato, finely chopped parsley or watercress.

Plain cake or biscuit.

½ pint milk, or a little less if he has had plenty of milk pudding and does not want it. But he should be offered ½ pint.

Piece of raw apple at the end of any or all meals.

Fruit with any meal.

ALL CHANGES, ALL ADDITIONS, MUST BE DONE GRADUALLY

Preferably, children should NOT eat between meals, but if they must, it should be some time before the next meal or their appetites will be spoilt. If they want 'elevenses' for example, fresh fruit or a fruit drink will be best. Sweets can be given after meals but the one-year-old baby should not yet know much about sweets.

Cleanliness and care of food is still important

So long as bottles are being used they should still be sterilized in the same way. As he begins to use ordinary utensils it is enough to see that they are kept scrupulously clean, well washed, well rinsed, dried with a clean towel and kept in a clean place.

It is difficult to state exactly when baby's milk and water need no longer be boiled. On this point it would be best to be guided by the expert who has known the child from birth and who knows local conditions.

The chief reason for boiling milk and water is because infants develop intestinal infections easily during the first year of life and only gradually build up a resistance to this during the second year. Also there may be germs in the milk or water, or germs may be introduced while the drink is being prepared.

There should be no relaxing of personal cleanliness when dealing with baby's food, even as he gets older. What he is given must always be fresh, properly cooked and attractively served.

When to stop giving sugar in his milk

As he gets the necessary sugar in his cereal and puddings he will cease to need it added to his milk. He might not like to be offered plain milk suddenly after always having had it sweetened, so it is best to cut down the amount of sugar gradually so that he will not notice, until he is happy without any at all.

If at any time he goes off his meals for a day or two add the sugar again to his milk, until his appetite comes back.

Travelling

Travelling with a child should be well planned if it is not to prove exhausting for all concerned.

It will be easy with the small baby who can be cared for in his usual way, and is likely to enjoy the movement. Any food needed during the journey for a bottle-fed baby or one in the stages of weaning, must be carried, ready for him.

Toddlers and older children need a light but nourishing meal before they leave, with little or no fat. Tempting picnic meals will be best, for long waits in restaurants which may then only supply unsuitable food, are to be avoided. Chocolates and rich foods should not be given. Snacks in between meals should consist of plain biscuits, apples, barley-sugar or boiled sweets, and not many of these. Nibbling while travelling should not be offered to prevent boredom.

Instead, games may be played to pass the time away happily, e.g. simple games such as 'I spy with my little eye something the colour of green (or beginning with "g"). What is it?' Singing in a car (or train if it does not disturb others) has great appeal. Lists of objects could have been made or pictures of them given, so that the child can tick them off as he sees them in passing (a brown cow, a black sheep, a yellow door, a dog on a lead, a copper beech, a dandelion, etc.). Nursery rhymes may be said, the alphabet repeated.

A period of rest should be insisted on at some time so that there can also be relaxation from the games.

Safety must never be forgotten. The child must sit quietly in such a position that he cannot roll about and hurt himself or interfere with the driver of a car. Train doors or open windows must be strictly out of bounds.

If the child is restless during car travel a break can easily be made so that he can have a short romp.

Supplement

Signs of readiness for weaning

As has been said there is no definite age or weight for weaning but baby himself gives plain indications if you know what to look for.

He may be satisfied with a moderate amount of sucking, reject the breast or bottle impatiently as soon as he has got his feed. He may appear to be hungry or dissatisfied and taste eagerly anything offered to him in a spoon or cup. Thus he shows he is ready to begin to be weaned.

On the other hand he may obviously love sucking and only be taken away from breast or bottle reluctantly. He may also show his continued need to suck by sucking fingers and thumb to a marked degree. He may be highly suspicious of anything in a cup or spoon, pushing away the hand that offers it, yet grabbing eagerly at breast or bottle. It would seem that such a baby is not yet ready for weaning whatever his weight and age. And even though he needs solid food and is persuaded to take it, he should still be allowed the breast or bottle as well, to satisfy the need he still shows.

Companionship rather than too much direct attention

A child who is regularly played with all the time he is not asleep can become a nuisance to other people and will not have the chance to develop naturally and to learn from his own experiments. But he need not feel deserted and alone. The sensible mother will offer a great deal of companionship but only a certain amount of playing, cuddling, nursing and petting.

Spoiling

Though most people agree that a spoilt child is not happier in the long run and is often disliked, it is not always easy to know when conscientious mothering stops and spoiling begins. It is a matter which needs careful thought. Sometimes a child becomes spoilt because his mother dotes on him so much that she cannot resist continuous petting and giving in to his whims. He may be very attractive in his ways, at least to her, and she may find him quite irresistible. Though a child at this early age cannot be said truly to be consciously calculating and cunning, he does somehow, to some extent, realize that he has got his mother exactly where he wants her and takes advantage of it.

On the other hand the child may simply have been fretful, difficult to manage, one who does not develop to the pattern suggested as 'average'. If his mother is anxious and worried she may spoil him in desperation, with the best of intentions.

The answer is to know as far as is possible by study and clear thinking what is best for the child, and to stick to it, secure in the knowledge that it is for his own good. When it is remembered that a child cannot possibly be expected to know for himself what is good for him then it is easier to resist spoiling. Those who talk glibly of allowing one's instincts to be the guide may forget that in all nature, parents guide, chastise and teach their young. They do not do so by giving in in every respect.

The more a child is given in to, the more demanding he will become. If he has been spoilt in early infancy, now is the time to undo any damage done, for with his growing intelligence he can be kept occupied with things of interest and can be made to understand that his mother cannot always be playing with him. He may not know the exact meaning of what is said but he will soon learn the meaning of such words as 'Play', 'Wait', 'Not yet', 'Mummy is busy', 'Mummy will come soon now.' His

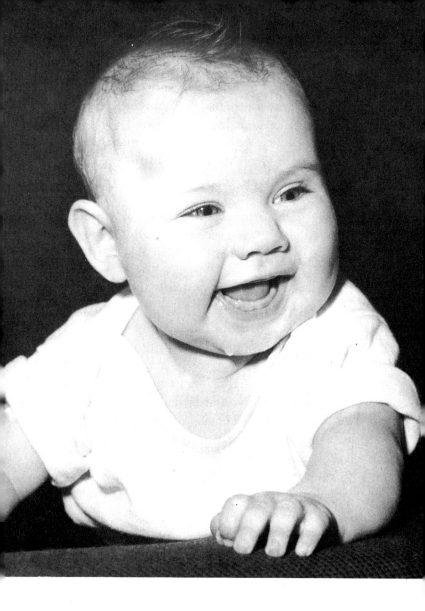

Playing on the floor

attention may be diverted from his own fussing if he is given plenty of opportunity to amuse himself suitably and attractively. He must start learning to be reasonable and independent, for the older he becomes before learning this, the harder a lesson it will be for him.

What the difference is between mothering and spoiling a small child, is often difficult to decide.

Clinging to mummy

The very young baby will be content to be handled by anyone so long as his needs are satisfied, but when he is able to recognize one person from another it will be an entirely different matter. He is beginning to move towards physical independence in such things as crawling, even walking and helping himself to certain things. But his emotional dependence is bound to be still firmly fixed on his mother. He cannot do without her. He needs to trust her in every way, to be sure of her continued presence. When she leaves him for short periods she must see that he is with someone he knows well, who will be kind to him. She should take it as natural and right that he will object to being left. This is a time when he is very sensitive, unsure and puzzled by anything strange.

If he is very upset when his mother returns from a short absence this is a time when a little extra spoiling will be good, until he is reassured and feels safe again. Clinging to mummy is not naughtiness, to be ignored or scorned, it must be dealt with tenderly and with gentle understanding.

Crying

What was said about small babies crying, still applies to a certain extent, but as well as making sure he is clean and comfortable, there are other factors to be considered. By now it really will start bad habits if he is picked up the moment he starts to cry (unless for a good reason, if he has hurt himself or is frightened). If, for example, he is brought downstairs to join the family in the evenings if he does not want to sleep, he will soon learn that it is easy to have a jolly evening instead of trying to get the sleep he needs. This would be starting a very bad habit which will be troublesome to stop.

If he is crying during play or at other times then it may be that he is bored. The answer is not to pick him up continuously, and spoil him, but to arrange matters so that he is occupied and contented. For instance if he is not happy in his pram in the garden and simply throws all his toys away, he might be consoled by being moved to a different spot, preferably with something interesting for him to watch. It might be possible to take him for a walk, or if the day is warm and sunny to take him out of his pram and try him in his play-pen in the garden, well wrapped up.

When a healthy, well-cared-for child of this age is not able to occupy himself contentedly in the surroundings offered him, then it is far better to change what is offered, rather than desperately petting and cuddling him.

CHAPTER 5

Children up to school age

During the second year of life. Temper tantrums. Suggested day for a toddler. Food. Rest. Helping him to become capable and independent. Time to learn to be sociable, to mix with others. Getting him ready for school. Animal pets. SUPPLEMENT. *More about toddlers. Examples of the different impressions a mother can make. When to offer choice and explanations. Curiosity. Imagination. Father's part.*

During the second year of life

Babyhood recedes and a distinct little person emerges. Physical progress will mean that he learns to walk well, to run, to feed himself, to make attempts at many things which have previously had to be done for him.

His growing intelligence will be marked. Not only will he understand much of what is said but he will begin to say quite a lot himself, even stringing simple words together.

Emotionally though, this can be a very tricky time. For he has to experiment, to learn so much, and to find out how to make use of and to control his growing powers.

Such emotions as pleasure, joy, laughter, fun, liking and disliking, annoyance, fury, shame, fear, worry, jealousy—these all appear. Naturally he shows his feelings and very often most emphatically and in a determined and obstinate fashion. Learning control of the emotions can be hard. His mother will need to show great patience and understanding and it will be a help if others who have anything to do with him can co-operate.

Then, too, it is a time of many frustrations. He longs to do something, tries hard and fails. He would like to tell someone about it but is incapable of enough coherent speech. He is constantly being misunderstood, or so it seems to him.

Temper tantrums

The tiny baby rouses deep protective love, the mischievous toddler can arouse emotions we never suspected we possessed! Toddlers cannot have their own way in everything but how to make this plain to them, with kindness and love, is not easy.

What we call bad behaviour should not throw the toddler into the limelight. If it does, he is likely to repeat whatever produced such pleasing results. Mothers and other adults often react to naughtiness in exciting and interesting ways. We all like attention, whatever age we are, but small children have an enormous need for it. If they do not get it by fair means then they get it by foul. So first they must have all the love and attention they need, as we have stressed so often, but unruly behaviour should not be a way in which they can count on getting more.

It is better to make a fuss over what we call good behaviour and aim at not fussing over bad behaviour. If a child kicks, yells, refuses to co-operate it is best to show no signs of being affected by it, to leave him alone till he recovers. If a battle starts he is more likely to repeat his struggles another time, for at least, even if he lost the battle, it will have interested and entertained him.

It is often possible to prevent tantrums. As well as seeing that he has a chance to play freely in a safe area, with plenty to interest him, tact must be used. There is no need to rouse any rebellious feelings unnecessarily, to interrupt him without warning, to allow him to be where he can see many things he wants and cannot have, to expect him to do things beyond his capabilities, or to meet his tantrum with another of one's own. He can sometimes be distracted from whatever is likely to make him blow up into temper, his mind and attention turned to pleasant things. He will often do a thing if it is simply expected of him and taken for granted, when, if he is offered a choice, or asked, he will refuse to do it.

Suggested day for a toddler

Drink of fruit juice on wakening.
Wash, brush teeth, dress, brush hair.
Let him play indoors or out in suitable weather.
Breakfast (more or less the same as the rest of the family).
Lavatory, wash hands, and mouth.
Play.

Taken for a walk at some time if at all possible, with pram or push-chair to rest his legs most of the time.

Rest before and/or after his dinner.

Wash hands. Dinner with the family.

Rest if he hasn't had one. (Sleep.)

Play or taken for another walk.

Then either an early afternoon drink of fruit juice and then a high tea or tea-supper, later, or tea some time between 4 and 5 p.m. and a drink of milk before going to sleep if he likes.

At some time he should have a game with his mother and/or father as distinct from play on his own.

Bath. Brush teeth.

Bed. Nursery rhymes or songs for a few moments, developing into bedtime stories as he can enjoy them.

Happily drowsy, contented, secure in his family's affection and care—

Sleep.

Food

After the process of weaning is completed, the next thing is to let him come just as gradually, to eat, digest and enjoy the well-balanced, carefully prepared meals which the rest of the family have. He will be likely to want what he sees others eating though he can be allowed to cling to any particular baby foods he really loves, for a time.

Note might be taken of a few foods which are not particularly valuable to him and might be undesirable at this age. These are not necessarily harmful, on occasion, but if taken too often have the effect of spoiling his appetite for the things he should have and at the same time satisfying his wish for food without giving him the food values he should get. These are cakes, biscuits, sweets, chocolate, pastries, ice-cream. If he does have these they should only be offered *after* he has eaten the more valuable food he needs, and then not too frequently.

Snacks between meals are still undesirable. Fruit juice or an apple would be best.

Let him feed himself if he has not already done so. It will be messy at first and he will need help and encouragement. But he must not play with his food. He should either be doing his best to manage it himself, or be allowing his mother to help or else it should be taken away from him. Yet meal-times must always be pleasant and happy. Whatever has to be done can be carried out with patience and good humour.

Rest

He ought to accept naturally whatever periods of rest and sleep during the day, are still offered to him. Observation of him and his behaviour will be the best guide. If he sleeps well when offered a morning rest, let him rest then. If he prefers it in the afternoon, or a short one in the morning and again in the afternoon, then let him do what suits him.

As the child leaves his third birthday behind, and goes towards his fourth and then his fifth birthday, he is often simply referred to as the pre-school child.

Doing his best

His personality will be changing incessantly and those who guide him are responsible for the direction which it takes, in spite of the fact that each child is so much an individual with a will of his own. So he still has to be studied and understood and not just left to grow up 'by chance' so to speak.

His physical development will be noticeable in that he can now co-ordinate his movements and control them. He will be able to express himself and soon to speak clearly enough for anyone to understand. He will manage to feed, dress and wash (with supervision at first). Play will be even more enjoyable for him now that he can do more of the things he wants to do.

Helping him to become capable and independent

When he wants to do things for himself he should be allowed to try, helped if necessary, encouraged, and praised. If he wants to help in the house appreciation can be shown. Every effort should be made to indicate that he is expected to do as much as possible for himself, to care for his own clothes, possessions, even his own room or cupboards. This is a time when children usually enjoy being allowed to do such things and advantage can be taken of this to instil good habits.

Time to learn to be sociable, to mix with others

Though he may have been accustomed to seeing other babies and opportunities may have been given him to play with little ones he is not likely to have shown signs of wanting to make friends much before the age of three or so, perhaps not even then. But he will, soon now, be ready to play with others. With more confidence in himself, more strength and control in his body, more understanding, he can learn what it means to be sociable.

Getting him ready for school

Up to the time when he has to go to school, the home, his parents and friends have created the only world he knew. Now he is to enter a strange new world, no longer controlled by his mother. Her part is to prepare him for it so well that he can take it in his stride, and to continue to provide for him the best home background she can.

By the time he goes to school he should have given up baby talk and be able to make his simple wants clearly understood in plain English. He should be able to attend to himself capably as regards going to the lavatory, washing face and hands, blowing his nose, dressing himself and looking after his own property.

If he has proved troublesome about being left with anyone other than his family, he should at least have been given some experience of this before the day comes when he is abandoned completely by his mother for what seems long hours to him. If short, pleasant separations from his mother have been a natural part of his life from time to time, this one will not prove to be too great a shock. Though his mother is of great importance to him it will not be unbearable for him to go to school if he has already realized that there are times when they have to be parted and is certain that they will always come together again.

He should know his name and address, have a good knowledge of road safety and have learnt not to panic but to turn to suitable people for help if he needs it. He should not immediately be expected or allowed to go or return on his own, so that teaching such things as avoidance of strangers need not yet be too deeply instilled into him.

There are a few parents who *like* their children to remain babyish as long as possible, who *like* them to cry when they are left anywhere. How much more satisfactory surely, to take a pride in knowing that the training, confidence and experience has been given to him, so that he can take his place happily and confidently in the world outside his home, while still relying on it deeply.

Animal pets

The decision to keep a family pet must never be taken lightly though it is good for the child to have one. Several points have to be considered:

1. Who will be completely responsible for the care of the pet? (The child should be taught to help.)
2. How much time can be spared to see that it receives *proper* care?
3. Is space available?
4. What pet would suit this family best?
5. Can arrangements be made during holidays?

The R.S.P.C.A. pamphlets on the care of animals should be read before any pet is brought into the household.

Pets must be kept healthy and well cared for in every way for the sake of the child's own health and also as part of his education. Sick animals should not be with children, for many reasons.

Naturally any pet chosen would be one which was considered suitable to be with a child. Yet even a good-tempered animal may harm a child in self-defence so it is important to teach the child to be kind and gentle.

Supplement

More about toddlers

Management of toddlers is not easy and in considering how best to deal with them we should remember the following points:

The toddler advances in some way, daily. What he can do and will do today and tomorrow, will be progress from what he has been doing previously. So those caring for him have to anticipate that he will be trying out something new all the time, and be prepared for it.

No one can ever know a child so well, even their own, that they can be sure that he would never do this, that or the other. Not much in this world is predictable, a child least of all.

Some sort of routine will be advisable so that all his needs can be met, e.g. times of fully supervised freedom and times of confinement in a safe place while his mother does her essential chores.

He needs to be handled tactfully. If he shows a distaste for his toys temporarily, he could be suitably dressed and given sand, water, dough, old jars, bottles, boxes. If he is being really tricky then mother can try to be too, e.g. instead of giving him things and telling him to play, she could simply leave them where they will catch his eye, yet be a little difficult to reach. He could be ignored while he enjoys helping himself. There is no need for interference in things which do not matter, and it can be reserved for the times when it does. For unwarranted interference can make a small child simply furious.

He should be dealt with firmly though gently but not hurried or confused or frightened. Though he must not be spoilt or given his own way in everything neither must he be nagged at, forced or jumped upon.

Examples of the different impressions a mother can make

When the toddler screams and kicks and throws himself around, if his mother rushes to him at once, to pacify him, almost as frantic as he is, what will he think? If on the other hand, he finds that tantrums of any sort simply result in his being ignored, he will learn, sooner or later, that there are better ways of gaining attention.

Maybe he has made a puddle on the floor, instead of asking for his pot, though he understands that he should.

If Mummy has stopped what she was doing immediately, to fuss and scold, then this may have been more enjoyable than being ignored or plunked crossly (because Mummy was busy) on to a cold pot. But the quiet tidying up of any mess and changing of pants in cold silence would have been effective, especially if there have been other times when he has been clever and asked for the pot and experienced the cosiness of a quick cuddle and a word of praise.

Perhaps others have treated his possessions without due respect. How should he then be expected to care for them, to continue to be satisfied by them? And further, how should he learn to treat the possessions of others with care? His toys should not be thrust hurriedly into cupboard or box or stuffed somewhere out of the way, if we want to teach him to look after things. There can be a great deal of contempt which will impress him, in the words 'Play with your *Toys*' often said in exasperation.

This is an age when the child is learning unconsciously from impressions made on him and by imitation, and by observing others. What he has before him to copy is of tremendous importance. The attitudes as well as the actions of the grown-ups he loves will undoubtedly be the guiding factors in his life.

When to offer choice and explanations

Most children under three find life easier if they are not asked to make up their own minds about things or to listen to explanations. It is best simply to assume that they will fall in with the plans made for them. For instance the child should not be asked whether he would like this, that or the other to eat or drink. A suitable diet should be provided. That he will consume what he is given should be taken for granted. He should not be asked if he wants to go for a walk or to bed, but simply taken there (after due warning if he is 'busy' of course). This is all part of a sensible taken-for-granted attitude to certain things which need to happen. They should be done without confusing him.

Neither will he be impressed at this age by being told that he must go for a walk to get fresh air so that he will grow big and strong, or that he must go to bed because if he does not he will be tired tomorrow.

The time for giving explanations comes soon enough. When the child begins to ask 'why' we must attempt to answer him, until then it is best not to worry him with reasons. When we do, they must be simple and within his comprehension.

Curiosity

From three years old a child's curiosity is even more intense. He is not only inquisitive in a baby way but inquires deeply and should be satisfied. With proper assistance he needs to be allowed to look at, touch and explore. His questions should be answered patiently. Father's help will be invaluable.

Imagination

At this stage too, the child will have a vivid imagination. His 'pretend' games can keep him very happy and it will be good for others to join in with them from time to time. Opportunities must be taken to point out that there is a difference between pretence and reality and that while we can all enjoy imaginary things they must never be confused with reality.

Father's part

Even a man who fights shy of a small baby and cannot be bothered with a toddler, may enjoy the companionship of the child as he grows more sensible and controlled. Certainly if the father spends time with his children it will be good for all concerned. Yet here too, a little thought and planning is needed. The tired father who comes home may be unwilling to play with or talk to his children unless it is understood that he, too, needs consideration. It will always be better for him to spend a short and pleasant time with them, than a long time during which he is irritable through over-tiredness. But now, even more than when they were smaller, the father can play a great part in their upbringing. His influence can be tremendous and if he does not exert it now, in a proper fashion, he should not expect suddenly, at a later age to be able to control his children.

CHAPTER 6

Safety and First Aid

In the home. Preventing burns or scalds. Preventing dangerous falls. Preventing cutting or piercing. Preventing swallowing objects, or poisons. Preventing suffocation. Preventing accidents from electricity or gas. Safe area of play. Out of doors. In the garden. In the street. In cars. On holiday. First aid. For burns and scalds. Falls. Head injuries. For cuts, stabs, abrasions, wherever there is an open wound. When objects have been swallowed. Poisons. Lesser troubles which may need first aid.
SUPPLEMENT. *Accidents. More about poisons. Fractures.*

In the home

It should not be possible for a serious accident to happen to a child in the security of his own good home. He must never be left alone even for a few moments without these questions being asked: Is there anything here which could possibly harm him in any way? Can he get out of this safe area by any means, and reach anything dangerous?

Accidents have happened in a matter of seconds—simply while the 'mother's back was turned' as they say. So that, the small child, unaware of the meaning of danger, needs to be guarded with the utmost care.

Making the home safe is something a father might do.

The following points are important:

Preventing burns or scalds

Open fires must have strong, well-constructed guards of a modern type. They must be securely hooked so that there is *no* possibility of a child being able to move them and there should be no opening he can manipulate.

Other fires may be fixed on a wall out of the child's reach. Modern electric- and gas-fires are all fitted with protectors but may also need to be guarded in the same way as an open fire.

There should not be a mirror above a fireplace or anything on the mantelpiece which might tempt a child to climb up to it.

Fires should never be drawn up with newspaper or have petrol or paraffin poured on when they are slow to burn.

Clothes and toys which are inflammable should be avoided. Anything which is inflammable must be kept in a safe place.

It should not be possible for a child to play with matches, candles, gas-taps, electric-plugs or loose leads.

Hot liquids must be kept out of reach. For additional safety, pan-handles and kettle-spouts should be turned away from the front of cookers. Teapots and hot-water-jugs, though safely on a table, may be dangerous if the child can pull them down by dragging at a tablecloth. Mats would be safer.

Hot-water-bottles must be put in thick covers and should not be filled with too hot water.

Hot liquids must not be passed over a child's head.
Cold water should be put in the bath first.

Preventing dangerous falls

Windows should be guarded with vertical bars.

Staircases need a gate top and bottom, the steps must not be polished or slippery, any covering must be in good repair and firmly fixed, with no mats at top or bottom which could slip.

Floors are dangerous if too highly polished. Non-slip polish should be used and mats should have safety corners.

Tidiness is important, so that there are no objects lying around which could trip anyone up.

Footwear must be in good condition always.

Anything spilt on a floor must be wiped up at once.

Chairs, cots, prams must be safe. Harness must be used.

Furniture must be so placed and constructed that it cannot be pulled over.

It should not be possible for the child to climb, unsupervised, to a great height.

Preventing cutting or piercing

Knives, scissors, tools, needles, pins, used tins, broken glass or china, razor-blades, dangerous toys, or anything which could possibly harm the child MUST NOT be left where there is any chance of the child getting at them.

Preventing swallowing objects, or poisons

NOTHING tempting or dangerous must be in reach. The eyes of a toy, if loose enough, may be swallowed for

Windows should be guarded

instance. Pills may look like sweets, medicines may look like something nice to drink, dangerous cleaning preparations may become even more dangerous if poured into, e.g. lemonade bottle.

Preventing suffocation

Baby is safest in his own cot where there is no danger of his being suffocated by a parent lying on him. The cot must not have deep soft pillows, the mattress must be firm, and bedclothes tucked in securely. Pets must not sleep in cots or prams, plastic bibs must be removed. Suffocation can be caused by plastic bags and these must be kept out of the way.

Suffocation can also be caused by such things as swallowing dummies, fluff from toys, blankets, clothes, choking when left to feed from the bottle alone, small baby being put to lie on his back and choking on vomit.

Preventing accidents from electricity or gas

The best advice must be asked for and followed, from gas and electricity companies. There must be no amateur repairs, faulty or unreliable equipment. Gas-taps and electric-points should be out of reach or protected.

Safe area of play

A safe area of play for a small child is ESSENTIAL. This can be a play-pen, a play-corner or part of room securely railed off, a room, several rooms and a landing, in fact any area anywhere, of any sort, *provided that it is completely safe in every respect*.

In this area it should be quite impossible for the child to be harmed in any way. He can be free to play in safety without close supervision though someone would be keeping an eye on him.

Out of doors

In the garden

Babies must be securely strapped in their prams and it should not be possible for the pram to tip up.

Toddlers must either be strictly supervised or temporarily confined to a play-pen, unless it has been possible for the garden to be made into a completely safe area.

Fencing must be in good condition, gates securely fastened so that the child cannot get out.

Within this confined garden there should be nothing available to the child which could do serious damage. Tools, gardening implements, weed-killer, gardening materials of all sorts, must be put away in a shed well out of reach. Only a tidy garden will be a safe playground for a child, and even then, there must be supervision.

The danger of water must not be forgotten. Not only lakes and rivers, but water-butts, shallow fish-ponds or swimming-pools, can be the cause of drowning. These must be efficiently guarded.

In the street

It should not be possible for the small child to get into the street. Doors and gates must never be left open even for a moment. Children should not play in the street.

When being taken for a walk the child must be completely under the control of the adult in charge. Sudden dashing after a ball or to look at something interesting should not be permitted.

At the earliest possible age the child should be taught road drill.

A good example must be shown at all times.

In cars

Small children should be held or watched by an adult who is not driving and consideration given to a safety seat or belt. They must be trained to sit still and not to be noisy, for their sudden movements or shouts could easily cause an accident. Heads, arms, hands, must not be allowed to stray out of the window. Passengers in a car, whatever the age, must not behave in any manner which could interfere with the driver or take his mind off safe driving. Young children should not be left alone in stationary cars.

On holiday

All that has been said about safety in the home applies to anywhere the child is living. Where it is impossible to make the necessary safeguards there must be continuous close supervision both indoors and out of doors. If freedom from responsibility is required by parents for short periods the child must be left with people who will look after him well. Special dangers of seaside and countryside must be noted. Gravel pits, rocks, cliffs, deep pools, tides—all must be remembered.

First aid

This is really a subject which should be studied separately, and completely, by all, and most especially by those who are to handle children.

Though accidents should not happen, they sometimes do, and it might even be admitted that they cannot always be prevented.

For burns and scalds

A *slight* burn or scald which has chiefly frightened the child and done little damage, may be held under the cold tap and dusted with bicarbonate of soda, perhaps bandaged lightly. The child may simply need comforting and some sort of pretence at treatment which will convince him that it will soon be better.

For everything, even slightly serious, the wrong treatment can be most harmful.

If clothing is on fire the child must be rolled firmly in rug, blanket, mat or a coat, and the flames extinguished.

1. He must be treated for shock. Shock shows itself by the child looking pale and feeble, breathing poorly, sweating, and when the pulse is feeble. The most important thing is to soothe and comfort him, to keep him warm and loosen his clothing. He needs fresh air without draughts. A warm sweet drink may be offered if there is anyone to get it for him but he should not be left alone while his greatest need is for comforting and assurance.

2. The doctor must be sent for or the child taken to hospital, whichever is the quickest.
3. Direct treatment on the burn or scald is difficult to lay down, opinions vary and change all the time. Those in charge of children should keep up to date and be aware of what their own doctor would advise. To do nothing directly to the affected part is better than to do the wrong thing. A covering of clean dry dressing such as gauze, lint, or clean cloth may be used while awaiting help.

Falls

Falls may cause bruises, sprains, broken bones, dislocations and head injuries.

Bruises in themselves are not likely to be serious and unless other damage is suspected may simply be given a cold compress, i.e. bandaged with a cloth soaked in cold water and wrung out, or pieces of lint soaked in equal parts of spirit and water may be applied.

Sprains may also be treated with a cold compress and the injured part supported while a doctor is being fetched. There will be pain on trying to move or if touched, and swelling, and discoloration.

Broken bones must be suspected when, as well as swelling and discoloration, the limb is in an unnatural position and cannot be moved, and where there is pain. The child should be got into the most comfortable position without moving him from the spot, and treated for shock. If he does have to be moved the injured part must be placed gently on a splint which can be made from wood, or thick paper folded several times, or cardboard. A bandage must not be tied over the broken part but well above and below it. The object is simply to keep the limb steady and to prevent further injury till medical help is available.

Dislocation is suspected when there seems to be deformity, discoloration, swelling, difficulty or inability to move, pain, numbness.

The child should be treated for shock as described previously and the damaged part supported in a comfortable position while someone gets a doctor.

Head injuries

If the child simply cries for ten or fifteen minutes, keeps a good colour and does not vomit, probably no harm has been done and all that is necessary is to observe him carefully for a time to make sure that he appears to be normal.

If, however, he loses colour and remains pale for any length of time, vomits, loses his appetite, shows signs of sleepiness or a headache, then the doctor must be called. In the meantime the child must be kept quiet and soothed.

Loss of consciousness (as in any other injury) would mean that a doctor must be found quickly.

For cuts, stabs, abrasions, wherever there is an open wound

Slight damage. The wound should be exposed, the hands of the person dealing with it well washed and scrubbed before the wound itself is well washed, cleaning away from the wound, not into it.

When clean it should be covered with a dressing or whatever seems necessary in the way of bandage.

Severe damage. The wound should be quickly covered with a firm pad or bandage and medical attention got at once.

When objects have been swallowed

A small object which has been swallowed without discomfort, such as a button, prune-stone, small coin, need not be a cause for great worry. The child must not be given a laxative but it is best to watch for the object within the next few days when nature will no doubt get rid of it.

More dangerous things such as safety-pins, pins, broken glass must be reported at once to the doctor, and no treatment attempted by the parent.

Poisons

Treatment for shock will be necessary while the doctor is sent for or the patient taken to hospital. Prompt action may be even more necessary here than in any other accident. The container that the poison was in and any vomit the child may have done should be taken to the doctor. There are so many different poisons, all requiring different treatments that only after close study would it be sensible to act without the doctor.

Whatever the accident was, a few general points should be remembered:

1. The child must be comforted and calmed. There must be no atmosphere of panic or fear. Whatever is felt must be hidden from the child so that he has confidence.

2. Cause of the accident removed if possible.
3. Treatment for shock given.
4. The child should not be moved if there is any doubt about the wisdom of this, but must be made comfortable.
5. *If the correct first aid is known it may be applied, otherwise it is best to do nothing and to wait for help.*
6. The meaning of the word 'first' aid must never be forgotten. The person who gives it must only give what the word implies—FIRST aid. It is for the doctor to do everything else in all except slight accidents.

Lesser troubles which may need first aid

Object in the nose. A child may push some small object up his nose. Care must be taken not to push this up further. He should be taken to the doctor if it does not come out easily.

Object in the ear. Again care must be taken not to push it in still further. If it cannot be removed very easily, then he must go to the doctor.

Sunburn. Plain cold cream, or one of the preparations bought from chemists should be applied and the area kept covered and completely protected from the sun till all redness has gone.

Animal bites (e.g. dog). Same treatment as for cuts, with a visit to the doctor.

Insect bites. Sting to be removed if it has been left in. Paste made with bicarbonate of soda and water applied.

Nosebleed. Often it is enough if the child is made to keep still for a short time. He should sit well up and have both nostrils compressed firmly with two forefingers for five minutes by the clock. If the bleeding does not stop within about a quarter of an hour, or comes back repeatedly, the doctor must be consulted.

Supplement

Accidents

The annual death rate from accidents is still very high. There are also many unnecessary injuries, severe and sometimes permanent, and millions of minor injuries.

It is interesting to consider the reasons for accidents in general, as well as the particular precautions just mentioned. They seem to be caused chiefly by carelessness, inexperience, stupidity, absent-mindedness, negligence— in short by not leading a well-planned life in a carefully made home. The kitchen and the stairs seem to be the sites of many accidents and they happen usually during the busy times of day.

More about poisons

It seems sensible if a child has swallowed poison, to make him vomit to get rid of it, yet there are a few poisons which are better to be treated in other ways. For example petrol would do more harm to the lungs by being breathed in, than it might do to the stomach so it is not wise to bring it up again when this might cause the lungs to be affected even more. A caustic or corrosive poison would burn the throat again when being brought up and might be better left in the stomach.

If, however, the poison is not one which will burn, or affect the lungs if breathed in, then it is best to make the child vomit.

This is done by giving him what is called an emetic. The back of the throat may be tickled and large drinks of tepid water given. Or two tablespoons of common salt (or one tablespoon of mustard) may be mixed in a glass of warm water.

With narcotic poisoning there may be convulsions, delirium or unconsciousness. In the case of convulsions or delirium the child must be kept still and quiet, where there is unconsciousness it is best to try to waken him or to keep him awake if he is drowsy. An emetic and drinks will be helpful in this case too, while waiting for the doctor.

It is important not to be misled into thinking there is no need to worry if a child appears to be all right after taking a poison. For some poisons have a delayed action (e.g. aspirin). The child must be closely observed for at least twelve hours, preferably longer.

Fractures

Children's bones are soft and are more likely to bend and splinter a little than to break right across. This is called a greenstick fracture. Other varieties of fracture are called simple, compound, complicated. In a simple fracture there is only slight injury to the surrounding tissues though the bone is broken. In a compound fracture, not only is the bone broken but the skin and tissues damaged, thus germs may enter and there is more danger. In a complicated fracture there is injury to some internal organ, some important blood-vessel or nerve, as well as the broken bone.

First aid should not be attempted in serious cases unless the subject has been studied fully.

Artificial respiration is best learnt in a first-aid class by actual practice. It must not be given if the child is breathing at all himself. But breathing may stop owing to drowning, suffocation, poisonous gases, electric shock and then it is necessary to give artificial respiration until help comes or until he breathes by himself again.

If drowning is the cause the child should be placed on a slight slope with his head downwards so that water will run out more easily. He should be face downwards, with hands above the head and the head turned sideways.

There are various methods which should be learnt and practised. As soon as he is able to breathe by himself, attention can be paid to improving his circulation and getting him warm. Breathing must be watched and further artificial respiration given if it seems to be failing again.

CHAPTER 7

Diet and health

Food values. Foods for growth, or body building. Foods for protection and maintaining health. Foods for energy and warmth. Teeth. Cleaning of teeth. Visits to dentists. When the teeth come. Eyes. A squint. Injuries or foreign bodies in the eye. Feet. Posture. Clothing. Winter clothing. Summer clothing. Night-clothes. Toilet training. Bed wetting. SUPPLEMENT. *More about toilet training. More about bed wetting. Vitamins. Calories. Digestive system.*

Food values

A well-balanced diet is one which supplies the foodstuffs needed to meet the many demands of the body. We have to provide for growth and energy, to protect and maintain health.

Foods for growth, or body building

These are called proteins and are found in meat, fish, poultry, rabbits, eggs, cheese, milk and to a lesser extent in lentils, peas, beans and a few other vegetables and fruits and certain cereals, bread and flour.

Foods for protection and maintaining health

These are called vitamins. They are to be found in oranges and other fresh fruit and vegetables: in liver, butter, whole-milk cheese, wholemeal bread, in fish oils, e.g. cod-liver-oil, and in small quantities of various other foods.

A full and varied diet should provide *mineral salts* which are also needed for growth and repair of body tissues, and for the regulation and working of every part of the body. The hardness of the bone and teeth depend on calcium and phosphorus. Iron and copper are needed for the blood-cells, iodine for the thyroid gland. These valuable minerals are found in all natural unrefined foods, e.g. whole grain, eggs, milk, fruit, vegetables and in meat and fish. Therefore what we have called the well-balanced diet, will provide the minerals needed.

Water is vitally important to the body though it is not nutritious. Most foods are composed partly of water and that is how the body's daily needs are chiefly catered for. (Drinks of water must always be offered to children.)

Roughage is contained in the fibres of vegetables, grain and fruit and will help to avoid constipation.

Foods for energy and warmth

These are carbohydrates and fats.

Carbohydrates are found in such things as sugar, flour and other cereals, potatoes, rice. (And of course in the things made from them such as bread, cakes, sweets, porridge).

Fats are found in milk, cream, butter, cheese, dripping, cod-liver-oil, fatty parts of meat or bacon, etc.

Very few foods contain only one of the necessary elements in a well-balanced diet, most are mixtures and contain several. To plan a sensible or well-balanced diet is not difficult, for it may be chosen from a very large number of foods.

The following is a guide to what a child should have in his daily diet.

1. Meats, liver, poultry, fish, rabbit or bacon.
2. Vegetables, greens or roots.
3. Starchy vegetables such as potatoes.
4. Fruit, at least some of it raw, and including orange.
5. Milk (in any form) $\frac{1}{2}$ to 1 pint, preferably approx. 1 litre.
6. Egg, cheese, soups.
7. Cereals, whole-grain bread and butter.
8. Puddings, sugar.
9. Water.

The value of knowing a little about the different and necessary elements in food will be found when a child goes off one particular item in the diet. Suppose, for example, that he does not want an egg, there is no need to worry if he is eating his meat, fish or other protein. Suppose he goes off meat, then fish, cheese or anything with protein can be offered until he is willing to eat meat again. Extra fruit may be substituted when he refuses his vegetables and vice versa. But it *is* essential to remember three important points:

1. That the body-building foods are well provided.
2. That energy-giving foods are available according to what the appetite dictates.
3. That protective foods are included.

Cooking changes foods in certain ways and the method in which this is done can add to its nutritive value. It can make it more digestible and nicer to taste and better to look at. It is important that the food chosen should be properly cooked and attractively served, so that the child will be tempted by it.

Teeth

The baby's first teeth are formed before he is born and this is why it is so important that the expectant mother's diet should be the proper one. His second, or permanent teeth begin to form a few months after he is born although they will not appear till he is six or seven years old. Therefore proper provision in the diet must be made, for the growth of good, strong teeth. Milk and cheese, vitamins and sunshine are particularly important.

Cleaning of teeth

This should be started as soon as there are any teeth large enough to be cleaned. The main purpose is to remove any particles of food which may cling to them, therefore cleaning should be after meals, the most important time being after the last meal so that they are clean for the long period at night. The teeth-cleaning routine can be made into a sort of game the child will enjoy.

Visits to dentists

From about three years old the child should be taken to the dentist. He will simply be asked to 'show' his teeth,

will get used to the whole business, so that if the time comes when he needs a small filling or any other treatment he will not be afraid. It will also ensure that any decay can be stopped before it becomes too bad.

When the teeth come

An average time for baby to start teething is at about six months old, though the first tooth may appear earlier or be delayed till much later. It can be quite normal (though not common) for a child to have no teeth till he is eleven or twelve months old.

The front teeth usually come first, then the back ones. They are likely to be all cut by the time the child is two to three years old. There will be twenty first or milk teeth.

Eyes

Eyes are so made that they keep themselves clean by blinking and crying and they do not generally need more than gentle wiping at bath-time. Should there be any discharge they should be bathed carefully and the doctor consulted if it is persistent or very bad.

A squint

It takes time for a new baby to focus his eyes together and there is nothing to worry about unless he continues to squint after he is about nine months old, when expert advice should be sought.

Injuries or foreign bodies in the eye

This needs treatment from the doctor. Eyes must not be touched inexpertly and the child must be prevented from rubbing.

Feet

Care of the feet cannot begin at too early an age. A baby's feet are soft and supple. His bedclothes should not restrict them into wrong positions and his first shoes are very important indeed. As he crawls, it is wise to watch that he is not repeatedly turning his feet outwards and his first walking steps should not be forced on him. He should walk only when he himself decides to do so.

A too-short sock or shoe will harm a child's foot and can seriously affect his health and happiness.

All good shoe shops have experts in the fitting of shoes and it will be best to consult one of these as to the right length, width, shape and flexibility.

Posture

Defects of posture are often preventable. Some of them are rather ugly. Prominent tummies, flat feet, knock-knees, hollow backs, flat chests, poking shoulder-blades and such imperfections are not attractive, apart from the fact that they indicate inferior health.

The main causes may be found in poor or inexpert care, generally e.g. too little rest and sleep in a bad position on a sagging mattress, or sleeping in a push-chair; unsuitable or inadequate feeding; lack of fresh air, sunshine, exercise; mouth breathing; tight, heavy or restricting clothes. Later, as the child grows older, other causes contribute to faulty posture, e.g. standing or sitting without change of position for too long at a time; sitting on high chairs with legs and feet dangling; wearing the wrong shoes; being made to walk too fast or being dragged along by a grown-up whose steps are bigger; defects of sight or hearing.

It would be well to consider these things though there should not be much to worry about if all is being done towards good general health. Remember that opportunities must be given for plenty of free supple movement.

Clothing

The points mentioned in Chapter 1 about clothing materials for the new baby should be remembered throughout childhood. The aim in clothing a child should be to keep him warm in cold weather and cool in hot weather. To burden him unnecessarily with too many clothes is as bad as not wrapping him up warmly. It is entirely a matter for common sense, taking into account the individual child, where he is to be playing, in what sort of weather.

As soon as the child is running about and leading an increasingly active life the chief points to consider are:

1. Hard-wearing materials, easy to launder.
2. Maximum amount of freedom of movement.

3. Material next to skin one which will absorb perspiration.
4. Light in weight, for heavy clothes are tiring and impede movements.
5. Easy to put on and take off so that the child can dress and undress himself.
6. Attractiveness.
7. Non-inflammable as much as possible.

From an early age the child should be taught to take reasonable care of his clothes, yet his play-clothes should be ones which allow him ample freedom in every way.

The use of rubber boots and macintoshes should be limited to rainy weather, for these, being non-porous, may mean that the child is wet through perspiration inside, while being kept dry from the outside. They are not good for very active play.

Winter clothing

Generally speaking a boy will need to wear a woollen vest and pants, flannel or corduroy knickers or trousers, a warm shirt (e.g. Viyella or Dayella), a woollen pullover and woollen socks.

A girl will need a woollen vest and knickers, a warm dress or a skirt and jumper, and woollen socks.

Out of doors a really warm coat will be necessary, and leggings for a small child. When really cold an extra woolly may need to be worn under the coat, also gloves and a hat.

Leather lace-shoes are best, strong and waterproof.

In the desire not to let the child be cold it must never be forgotten that he needs freedom of movement.

Summer clothing

A boy will need a cotton or Aertex vest and pants, cotton shirt and trousers.

A girl will need a thin vest and knickers and a summer dress in a light, cool material.

Both can wear thin socks and well-made sandals. A sunsuit may be enough in hot weather. A lightweight cardigan or pullover may be useful at times. Hats should be worn in very strong sun.

Night-clothes

Very warm materials used for pyjamas or night-dresses in the winter and cool cotton ones in the summer.

A dressing-gown and bedroom slippers will be needed.

Toilet training

There are different opinions as to when toilet training should begin and each mother may decide for herself which she prefers.

1. Baby can be held out on his pot, after feeds, from the time he is a month or six weeks old. This does not mean he is learning control yet, but simply getting into good habits and becoming used to a pot rather than a wet or dirty nappy.

2. He need not start training at all until he is about a year old, the theory being that by now he can understand and could not have been expected to do so sooner.

This second method has not the advantage of giving the baby an early expectation of cleanliness or habit. It means that a strange object (the pot) is suddenly presented to him when he is at an age to notice and possibly object strongly, instead of its being a regular part of his life.

For success in training the following points matter more than anything else and the time when training starts is of less importance:

1. It must be remembered that the child cannot be trained by any certain age. He can only be trained when he himself reaches the right stage of development. The age for this varies considerably.

2. Patience, pleasantness, lack of anxiety on the mother's part, encouragement, optimism, praise—are essential. There must never be disgust.

3. Scenes must be avoided. If they happen, they must be ended quickly and calmly and the child's mind turned to something else.

Bed wetting

Though good training in clean habits is helpful, this is an entirely separate problem, for the child does not intentionally wet his bed. On no account should he be scolded or upset for this will only make him worse. It is very usual for children under two to wet their beds at night and many children are much older before they are completely reliable.

The occasional wet bed after the age of two to three should be accepted in a matter-of-fact way. It may have been caused by overtiredness or some unusual happening. There may seem to be no explanation, and it is unimportant.

The frequent bed-wetter, over the age of say three, is more of a problem. The child himself may be worrying about it and the parents may find it hard not to show that they are upset, or even annoyed over such practical problems as the laundry. In such a case it would be best to ask for medical advice.

Generally speaking, bed wetting should be treated as casually as possible and never discussed in front of the child. He may be praised and encouraged when dry but not punished when wet.

Supplement

More about toilet training

That a child should learn clean habits is something which appears to assume enormous importance to the majority of mothers, and as such, needs to be given more attention than it does in fact deserve. Mothers often appear to feel great shame and disgrace when their toddlers do not take readily to the pot or lavatory. They hesitate to suffer the anxiety of taking a child out who is unreliable, and grumble at the washing they have to do. Very often, for such reasons, unnecessary problems arise.

Faulty training or lack of understanding on the mother's part has been known to create difficulties of many sorts, e.g. persistent refusal to use a pot, and to be wet and dirty by the age of three and even older, puzzling behaviour such as running away when they want to perform, screaming and other behaviour problems which may not appear to be directly associated with the pot but which well may be. Psychologists have traced, and later cured, such things as disobedience, resentment, obstinacy, neurotic behaviour of various sorts, to faulty toilet training.

Certain facts should be realized, and no amount of goodwill on the child's part can alter them:

1. The child is not capable of control until a certain stage of development has been reached by his body and nervous system. Whatever the training has been, it can make no difference if his system is not ready.
2. This stage of development may have a temporary setback or be definitely retarded owing to wrong handling. These functions are so closely connected with his nervous system that if something affects it, the normal development may be delayed. Therefore the effect of worry or unhappiness could easily be to slow up his ability to reach the desired control.
3. The child cannot be expected to share his mother's

views on the importance of being clean. It does not matter to him if he is wet or dirty for his mother will soon make him comfortable again. It is of no consequence if he performs in napkins, his clothes, the floor or the pot though of course praise from his mother when it is in the right place will encourage him. The point is, that he is unlikely to have instinctive feelings about it though some mothers seem to imagine he will. His attitude will only arise from what he is taught.

More about bed wetting

What has just been said about toilet training during the child's waking hours can be taken into consideration when thinking of bed-wetting difficulties. The size of his bladder, the stage of development of his nervous system, and his reactions to his mother's reactions will all play their part.

It is thought that bed wetting can occur during disturbing dreams of some kind, or if a child is upset in any way, perhaps jealous of a new baby. The mother can take stock of his daily life and see whether she can improve it in any way generally. In a very few cases there may be some physical cause but there are usually other symptoms which will help the doctor in his diagnosis. In most cases the bed wetting is due to some upset of the child's feelings or nervous system even though this is slight or unconscious. He may be upset by some change in his usual daily life, new surroundings or new people around him. He may want more attention. He may simply have had too much excitement.

Large drinks last thing at night should be avoided though it is not now thought that curtailing the amount of fluid given throughout the day achieves any better results. In fact he should have a normal fluid intake during the day.

It is quite a common practice to pick the child up, while asleep, to empty his bladder, about 10 or 11 p.m. This often helps and should be done quietly in a soft light, without disturbing him. Most children hardly stir and go off to sleep again at once.

When the older boy or girl, from five or six years onwards, continues to wet at night it may often be hard to go on being patient but still, obviously, scolding will not help. The child can be told that others have had the same problem and overcome it and that there is no doubt that he will too, in time. The older child could be expected to remove the wet things from his bed and put them into cold water to soak. This need not imply punishment, simply a reasonable contribution to the extra trouble he causes.

The only way to help him is to try to see that he is a happy and well-balanced child, with no worries which could be put right, with no cause for anxiety at home or at school.

Much depends on whether a child sleeps heavily or lightly and proper medical advice must be taken for any child who is a persistent bed-wetter.

Vitamins

The body needs frequent small amounts of the substances we call vitamins, so that it can keep in good working order.

Vitamin A is plentiful in milk, fat, egg yolk. Carotene is present in green and yellow vegetables and is converted by the body into vitamin A. This keeps part of the eye healthy, also the linings of the intestinal, urinary and bronchial systems.

Vitamin B Complex contains several different vitamins but they are found mostly in the same foods so there is no need to separate them in our minds for the purpose of planning children's diets. They are found in eggs, milk, whole grains, certain vegetables and fruit, and in liver and meat. Every tissue in the body needs these, if the nerves, skin and blood are to be healthy.

Vitamin C is mostly found in oranges, grapefruit, lemons, blackcurrants, tomatoes and raw cabbage, though smaller quantities of it are in other fruits and vegetables, e.g. potatoes. It is easily destroyed by cooking. Bones, teeth, blood-vessels, other tissues and most of the cells in the body, need Vitamin C. Lack of it in the diet makes the body less resistant to disease and may cause scurvy.

Vitamin D is in cod-liver-oil or any fish-liver oils. Grown-ups get enough from the small quantities present in butter, fish, eggs, etc., but children need the larger quantities as long as they are growing. Nowadays it is added to margarine. It is particularly good for the growth of the bones and teeth. It helps to get calcium and phosphorus absorbed into the blood and deposited into the growing parts of the bones. Lack of it may result in rickets.

Calories

The amount of energy the body uses and how much food is needed to meet its demands is measured in calories. If enough calories have not been taken then hunger is felt. No guide, other than the healthy child's own appetite, is really necessary, to the number required.

Some foods, high in calories, are the fats (butter, margarine, vegetable oil, etc.), sugars and syrups, meats, fish, poultry, eggs, cheese, grains, milk. There is, however, no need to study and count these for the normal healthy child so long as the proper well-balanced diet is given and care is taken not to overfeed.

Digestive system

Digestion of food begins first in the mouth. It should be well chewed and will mix with the saliva. Then in the stomach it mixes with gastric juices which continue the process. Finally, in the small intestine some of the nourishing part of the food passes into the blood and the rest into lymph-channels by means of which it finally enters the blood. It then travels to the different parts of the body through the blood-stream. The waste matter, of course, is removed by the intestines, when the bowels move.

CHAPTER 8

Sleep and general health

Fresh air. Sun. Appetite. Unwillingness to take milk. Crying. Crying for his mother. SUPPLEMENT. *Emotional problems. Aggressiveness. Jealousy.*

Sleep is a basic need at all ages and though it has been mentioned already, it is so important that it deserves separate study. Here is a guide to the amount of sleep needed. The newborn baby sleeps most of the time. He probably only wakes when he is feeding or being attended to.

From birth to 1 month he may sleep approximately 20–21 hours out of 24.

1–3 months—this will gradually decrease to 18–19 hours out of 24.

3–6 months—this will gradually decrease to 17–18 hours out of 24.

By about 1 year—this will gradually decrease to 13–15 hours out of 24.

By about 2 years—this will gradually decrease to 12–13 hours out of 24.

Right up to about 5 years about 11–12 hours out of 24.

It would be foolish to insist that at any age a definite amount of sleep is needed or taken, but it is always helpful to have an idea of what suits the majority of children.

The small baby, still having five feeds a day, will be unlikely to have more than seven to eight hours unbroken sleep at night, but as soon as he stops his late evening feed, up to the age of about one year, it is advisable for the night sleep to be an unbroken one of about twelve hours, and the other hours of sleep made up during the morning and/or afternoon.

It is very easy to allow baby to grow into a bad sleeper yet not at all difficult to help him to sleep well if thought is given to the matter. Here are some of the ways sleep can be encouraged.

1. The small baby needs gentle, careful yet firm handling so that he feels secure. He needs ease, peace, relaxation, yet there is no necessity for hush and unnatural quiet.
2. He should be *expected* to sleep. Something of this will be conveyed to him. If someone hovers round him, waiting to see if he will waken, he will soon begin to realize that it is easy for him to gain attention. Yet he must be where serious cries could be heard.
3. He should not be handled unnecessarily. The suggested timetables for his day should be studied. Though he may enjoy extra petting and cuddling, it will tire him, make him irritable, and he needs rest.

4. He must never be put down in cot or pram to sleep, unless it is perfectly certain that he is comfortable in every way. He must be clean, properly dressed and covered, well satisfied by his feed and should have got some wind up. In short he should feel well cared for and much loved.
5. A relaxed and confident person in charge influences the baby to be the same.
6. He should not know any feelings of hurry and worry. Yet lengthy lingering over a feed or getting his wind up is not necessary if the whole thing is attended to competently.
7. He needs plenty of fresh air.
8. If he seems wakeful and cries for more than five minutes or so, the cause should be investigated. It is best not to pick him up and carry him to wherever he spends his waking times, to cuddle and play with him —but rather to do whatever needs to be done, by his pram, or cot. In warm weather his napkin can be changed wherever he happens to be. If he is wet or dirty, a damp flannel and towel to clean him with should be brought. If he has wind, it can be got up and he should be returned soothingly but firmly with the expectation that he will now sleep. In cold weather his pram can be brought inside for these attentions or the window shut if he is indoors. A change of position may help. He will soon get the idea that if he does not sleep he will not be given an extra time of play though he *will* be attended to and made comfortable.

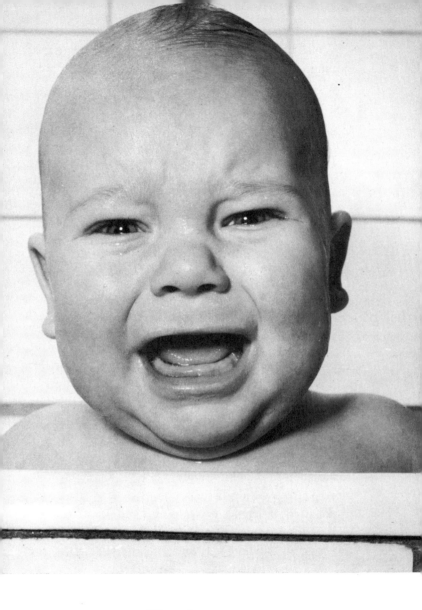

What is the matter?

While baby is small then, it is simply a matter of dealing with him competently, making the right impression on him and starting off a good habit. But as he gets older and needs less sleep it is still important that he gets *enough*. What has to be done now, is to give him *ample opportunity* for the sleep he needs, and he will take it.

This means that for the first few years of his life he will continue to be put to bed early, to be *in his bed* for about eleven or twelve hours each night. Having formed good habits in infancy he will take as much sleep as he needs out of that period. Then until he is about four or four and a half years old he should also have a rest period either in the morning or afternoon when he can take a nap if and when he wants it.

Handling at such times is still of tremendous importance. What has been learnt about helping the small baby to sleep well, still applies. No child should ever be expected to sleep or rest unless he is happy, comfortable and well cared for in every respect. But as the hours of sleep decrease his waking times should be filled suitably.

From the age of four, five or six months he will like to have toys to look at as he drifts off into sleep. He will love to have something to watch which interests him yet it should not excite him unduly. Often, for example, to have gently swaying branches of a tree to gaze at, may make all the difference between tears and peace. As soon as he is able to clutch a toy this will help; when he reaches the stage of cuddling a favourite he must be given one which appeals to him.

From about a year onwards, or less, he can always have a variety of toys to play with, to keep him company and to occupy him while he is being given the chance to take the sleep he may or may not need. He should know that his mother is near but should be taught to understand that he must not expect unnecessary attentions from her.

The suggestion that he only has a rest period in the daytime until he is four, or four and a half, is because he should be well accustomed to doing without it before he goes to school. It would not be good for him suddenly to have to forgo his usual daytime rest at the same time as he has to cope with a new and more strenuous life. In any case, like everything else, this is a change which is best made gradually. First of all what might have been a two-hour rest can be slowly shortened till it is only half an hour. Then he may be put to rest only every second day perhaps, or if he seems tired, until he ceases to take more than the long night's sleep which he will continue to need for some time yet.

Further ways to help the child to sleep, after infancy, are:

1. Suitably chosen bedtime stories.
2. Making certain that his activities during the day have been of the sort to make him healthily tired and ready for sleep.
3. Never letting him be unhappy at bedtimes, or using bed as a punishment, which might make it become a place associated with his childish misery.

4. Always giving him due warning so that he is not rushed into bed in the middle of a game. Trying to take him to bed when his day is beginning to be almost boring—not when some other excitement is going on which he does not want to miss.
5. Carrying the toddler affectionately, leading the older child by the hand, tucking him up without haste, carrying out his requests amiably, e.g. Teddy here, somebody else there, and so on. Seeing that everything is as he wants it before leaving him. What he needs to make him happy now is more than he needed as a small baby and he should not be left till he is satisfied. The assumption that he will be good and drop off to sleep will help. So will the knowledge that he will be attended to if genuinely necessary. He can be chided gently if he becomes demanding and parents should refuse to be bullied by him.

If, in spite of all efforts, the baby cries and does not sleep, it must be remembered that it is usual for babies to cry or even scream at times and certainly the best of babies will have wakeful periods. Therefore there is nothing abnormal in his behaviour and he must simply be treated in a calm fashion while a check is being made on all the things which should have been done for him. If he continues to cry incessantly, advice must be sought from doctor or welfare centre. Observation and common sense should make it possible to distinguish between genuine cries of distress due to wind, pain, discomfort of any sort, and the cries which are really just baby talking to himself, practising his voice, or 'trying it on'.

A toddler or older child may suddenly lose his good sleeping habits due to fear of the dark, bad dreams or some problem which has arisen for him. The parents' attitude in such cases must be nicely balanced. Sympathetic concern needs to be shown, yet there must not be over-anxiety and it must be assumed that whatever is wrong can and will be put right, and he can then get off to sleep.

Fresh air

Babies should be out of doors for at least two or three hours every day unless it is foggy. As they get older they should be out for longer periods and when indoors their rooms must be well ventilated. If, for any reason, baby cannot be put outside then he should be dressed and covered as if he were going out and put near to a wide-open window in a room which has just been cross-ventilated. (This means both window and door open for a time to allow complete change of air.)

At all times baby should be in a fresh atmosphere, never in a stuffy one, although windows need to be closed during bath- and feeding-times, in the cold weather. His window should always be open at night. Inexperienced people are naturally anxious that baby should not catch cold, but he will be more likely to do so, if he is kept short of fresh air and overheated.

For toddlers and older children *fresh air and exercise* can be considered together. The toddler may still have his daytime rest and sleep outside, but as he is likely to play for a time first, he must be exceptionally well wrapped up in cold weather and his mother can watch so that she can cover him up when he goes off to sleep. The house must be airy for him too, as indeed for all of us, but he will now benefit greatly from exercise in the fresh air. He will need walks and play in the park or garden but there must always be the opportunity for him to sit down if he gets overtired, on a seat or push-chair, until it is certain that he is old enough not to suffer from over-fatigue. Every child should be given all encouragement and help to be out in the fresh air for as long as seems right for his age and for the time of the year.

Sun

Ultra-violet rays in direct sunshine produce Vitamin D in the body tissue. Sunbathing is good for babies, from the time they weigh say 10 lb. or so, and for the rest of their lives. Yet the sun must be treated with respect. A baby will benefit from having his whole body exposed to a warm sun for a few minutes daily at first and never for more than ten minutes or so, although he may be left for longer with his legs in the sun provided he does not burn or blister.

Toddlers and older children will love, and benefit from, being allowed to run around in brief sunsuits whenever the weather permits. At first they too must become accustomed to it gradually, only being exposed completely for a short time and being covered with light porous clothes or kept in the shade the rest of the time.

At all ages hats must be worn in strong sun. Although sun is valuable, enjoyable and a great treat to us in this country, it must only be taken in moderation, especially at first.

Appetite

Whether or not a child has a good appetite depends largely on his health and the sort of life he leads. It can be very worrying when a child goes off his food. In order to avoid difficulties at mealtimes:

1. He must be leading a healthy life.
2. The right food, attractively presented, in suitable quantities and in variety, should be offered at reasonably regular times.
3. He should not be allowed to 'nibble' in between meals, thus blunting his appetite when meals are ready. When sweets are given he should have them after a meal.
4. He should not be overtired, upset or worried by anything.
5. Others eating with him should set a good example.
6. He must not be made to hurry over his food yet he must not be allowed to play with it. Reasonable time should be allowed for each mouthful, but any tendency to linger checked simply by clearing away the meal when he obviously does not want any more, and not offering anything further to eat until next mealtime.

7. It should be accepted that, like everyone else, he will have days when he is not hungry. This is perfectly natural. So he should never be forced to eat and no one should show undue concern if he does not want to, thus giving the matter (and him) undue importance. Coaxing, bribing, threatening, punishing are quite useless in the long run. The child must learn to regard food as something to satisfy his hunger, and to enjoy, not as something which will attract attention to himself and cause him to be the centre of a scene.
8. As he should not be over-fed, small helpings which he can be expected to finish, are best. He can be given a little more if this seems wise.
9. He should never overhear discussions about himself and his food. If his mother wants to discuss her problems she should not do it while he is present.

Unwillingness to take milk

After baby is weaned he may possibly at some stage show a distaste for milk. If this should happen it is best not to make a fuss or show concern though he ought to have at least a pint of milk a day. It is better to stop offering it as a drink for a time rather than make a fuss. There are other ways that he can be given milk which will do just as well, e.g. with cereals (pre-cooked and dry cereals absorb a lot of milk), porridge made with milk, milk puddings of all kinds, soups made with milk instead of stock or water, potatoes mashed with milk or baked in milk, fish steamed in milk and many other dishes.

Milky drinks can be offered instead of plain milk but very often when a child does not want milk he will refuse these too. Sometimes though, a new mug or cup may tempt him or he may decide to drink if he is given a little jug and allowed to pour out what he wants from time to time into his cup. While very young his hands could be guided. He may even be satisfied if the merest suspicion of weak tea is added to his milk or thrilled to be allowed to mix in, for himself, a small quantity of something like powdered chocolate or Ovaltine or one of the cereal and malt preparations. To be allowed to drink through a straw or transparent plastic tube could be tried also.

If milk in almost any form is refused then the deficiency can be made up by offering other foods with calcium in them, e.g. yoghourt, cheese. Cottage cheese has a low fat-content and is easily digested. Children often love all kinds of cheese, grated, spread in sandwiches or eaten in a piece. At the same time the rest of his diet must be reviewed to make sure that it is a proper one.

If he refuses milk in any shape or form for more than three or four weeks then it is best to consult doctor or clinic.

Crying

With the small baby it is chiefly a question of wondering why he is not sleeping, if he is crying, and what has been said about good sleeping habits should be examined.

But as the infant grows and his waking hours are longer, the matter becomes more complicated as was pointed out in Chapter 5. Sometimes the reason for tears is obvious, e.g. he needs comfort if he has fallen or to be ignored if he is simply crying for something he cannot have. But there are times when the child cries apparently for no reason or he may be ready to cry for the very slightest reason. Here are a few questions which could be asked in such cases, remembering too, all that was said in Chapters 4 and 5.

1. Does he have enough love and security? A good home background, no quarrelling parents.
2. Could he feel jealous and unwanted?
3. Are his physical needs properly met? e.g. food, exercise, fresh air, cleanliness, clothing, sleep, rest, etc.
4. Could he be over-stimulated? e.g. more expected of him than he is capable of, thrilling stories or television or too much excitement of any sort?
5. Could he be afraid of anything? The dark? A big dog? The unknown? Another child?
6. Is he bored? Has he enough to occupy himself with? Enough change? He cannot concentrate on one thing for too long.
7. Could he have bad dreams?
8. Is he being given too much choice? He needs leading and cannot always make his own decisions. He may need help if he is not to worry.
9. Has he enough companionship? Is he ever lonely?

It is usually easy to know when the child is crying purely from temper, almost all children show temper at times. They have wills and minds of their own and cannot be expected to submit to others without protest.

What caused the tantrum? Can it be avoided in future? If it was some matter of safety, health, discipline, then the child cannot have his own way. Even then could the matter be approached in a different manner, more acceptable to the child? It might be worth trying.

If it was not important then why insist?

Was the child justified in any way? Could the grown-up have been unreasonable or impatient or even ill-tempered? Did she nag maybe?

While the child is at play he needs an area where he is free to do as he pleases without constant interruption or being told 'Don't touch that.' He should not be stopped suddenly in the middle of play either, or expected to do anything beyond his capabilities.

Therefore in order to reduce crying and tantrums to the minimum, it is best to make as few rules as possible, to insist only when necessary, to use tact and to try to distract an angry child's attention rather than to prove to him grimly who is master.

Crying for his mother (see also in Chapter 4)

Anxiety in the small child that his mother is going to leave him cannot be entirely avoided. Nearly all children go through this phase to some extent though some are worse than others. To help the child over this period the mother should:

Try to understand. His whole world revolves round her.
Teach him early independence as far as is possible.
Give him trust and confidence in her, so that he can feel

not only that he can depend on *her*, but that any arrangements she makes for him will be all right.

Never to lie to him, e.g. if she should leave him saying she will be back soon and then disappearing for several days. At the same time remember that a child's conception of time is not accurate, although he can, at least with help, count off the nights of absence.

Accustom him to meeting and being with other people.

Only leave him with people he already knows and likes, people who will understand him and be kind.

Make the period of separation as short as it can be, especially at first. Keep in touch with him, e.g. a daily postcard. This will assure him that she has not forgotten him and will soon be with him again.

Expect and accept the fact that he may be upset by the separation and not make it worse by scolding, chiding or showing irritation. Instead, on leaving, express her own sadness at having to leave him (without emotion) and show great pleasure when she returns.

Supplement

Emotional problems

Aggressiveness

Very often children are aggressive because they are unable to express their feelings in words. They may be jealous, frustrated or troubled in some way. Or what appears to be aggression may simply be that they lack control over their actions or that they are curious and have to experiment in order to see what will happen.

It is not easy for parents to increase their affection for a child when his behaviour is being particularly irritating but if they can do so, and at any rate, remain calm, they will help the child to recover from his bad feelings. Whereas if the parents return aggressiveness for aggressiveness they are unlikely to get anywhere.

Jealousy

This can be very disturbing and one must accept the fact that this is a natural failing. If parents can convey to the child that they accept and understand his jealous *feelings* and sympathize with them, yet would not allow harmful jealous *actions* (such as hitting baby), they will be protecting the baby yet helping the child. He will also be learning self-control and learning a useful lesson.

When a second baby arrives the first child is often difficult and will make his jealousy obvious.

This might be prevented if:

The first child can be made to feel important and *secure* in his mother's love.

The father, and friends help by devoting more time to him, by not neglecting him and admiring the new baby too much in his presence. (They should remember that the new, tiny baby does not know what is going on yet and so cannot be hurt, while the toddler can.)

While the baby is being attended to, the toddler is kept *busy*, fully occupied.

Any changes which have to be made for the new baby's arrival, are made well beforehand, so that the toddler does not connect changes with the baby, e.g. if he has to move to a different bedroom, some of his things handed on to baby. He should not be asked to part with anything which he truly loves but could be asked to *give* anything his mother wants for the new baby. He should not be expected to adapt himself to great alterations in his life, surroundings and possessions, at the same time as he has to adjust to the presence of the new brother or sister.

The first time he sees his mother again after the birth it will be better if she does not have the new baby in her arms or even in the room. Let her greet him and give him her full attention and affection first and then ask him if he would like to see his new brother or sister. He must not be given the idea that someone else is *replacing* him in his mother's affections, but must know that someone is joining the family.

He should not see the new baby being given many presents while he gets none. Tactful visitors may bring him something too. He should be allowed to help his mother, e.g. fetch things for her, pass her things at bath-time, splash water on baby, sprinkle a little powder on baby. In the last few weeks before the new baby comes the mother must be careful that she does not make a greater fuss of the toddler than usual. This would be quite natural, for she may be feeling that this will be the end of a special time they have had together, she may be resting and have more time to cuddle and pet him. Instead, she could concentrate on teaching him to be as independent as is possible for his age, encouraging him and showing him how to play and occupy himself alone.

Finally, as the mother gets into a routine of preparing the bath for example, for the new baby, she could equally get into a routine of preparing something for the toddler to occupy himself with while she is busy. It will be easier for her in the end, than coping with a troublesome child. He could be equipped for messy play perhaps, given plasticine and told to make pounds and pounds of sausages, potatoes, oranges, apples, if he is not capable of anything else!

It can become an accepted fact that the toddler 'works' while Mummy is busy with baby. Mummy can inspect his work with great interest and encouragement and it can range from bed-making (never mind how badly), to dusting, building, drawing, doing something creative. *Jealousy between children as they get older* should not occur to any distressing degree if they are treated alike as far as possible, all treated fairly, and if each child is satisfied with his own share of love and attention, he will not be likely to grudge what others receive.

CHAPTER 9

Character training

The home. The mother. The father. Good habits. Bad habits. Truthfulness and lying. Why should a child lie? Obedience and discipline. Punishment. Good temper and self-control. Good manners. Fairness, unselfishness and thoughtfulness. What is meant by bringing up a child to be a good Christian. Religion. SUPPLEMENT. *Object of character training. Appreciation of the child's view.*

It is comparatively easy to feed, dress, wash and supply a child's physical needs satisfactorily. Knowing how to bring him up so that he will become a fine character with a sound mental outlook and strong morals is very much more difficult.

Unfortunately there is no simple recipe for forming character. Some advocate leniency, others strictness, yet others steer a middle course between the two. It is not possible to decide which method brings the greatest success. There are, all the same, several things which quite definitely have a tremendous influence.

Firstly (and this is obvious though too often disregarded) example and environment are of primary importance. The child is likely to be, largely, the product of his background. If his home and the example of those he loves in it, are truly good then we can anticipate that he will turn out well. Present-day research shows that childhood experiences play a large part in the sort of adult a person becomes though of course we all know of cases where the best training in what seemed to be a really good home turned out a poor type of grown-up. We also know of the valuable members of society who have had no decent early training.

Secondly, in dealings of any sort with the child, it is of the utmost importance that grown-ups should be consistent. How can he come to know right from wrong if he is scolded one day for some misdemeanour, and gets away with it the next? He must be able to know what is expected of him, in order to comply. It is helpful too, not only for the mother herself to be consistent but for the mother and father (and any other person who deals with him) to decide together how he should be guided. He is less likely to become confused in this way and will arrive sooner at the stage of being able to conform to acceptable conduct and to discern right from wrong.

Thirdly it must always be remembered that the child's stage of development must be taken into account when some particular kind of behaviour is expected from him. This is perfectly simple really, for the observant and thoughtful parent will know very well what the child is capable of and will not tax him too far.

Almost everything we do with a child is contributing to the formation of his character, directly or indirectly.

Parents influence their children every minute of the day, not only by their example but by what they do and say to the child. Each child has his individual tendencies and possibilities, but what these are made into is within the power of the parents to a marked degree. Although the formation of character is happening naturally, throughout a good upbringing, there are some aspects which can and should be deliberately planned.

The home

A dictionary definition of the word home is 'the place felt to be an habitual abode, the scene of domestic life with its emotional associations'. This in itself is enough to make us realize how important it must be to provide the *right* scene of domestic life, the *right* emotional associations, in the habitual abode.

Ideally there should be a strong atmosphere of love between the parents and from them to the children. They should work together to make a real home for their family, a place where they are cherished and understood. Children must be protected and cared for, made to feel that they are precious, that whatever happens in a cruel world they have a safe place, a dear, highly esteemed niche all their own. It would be easy if there were rules which could be laid down for making such a home but this is something which each family has to design for itself.

No child has asked to be born, and before he is brought

into the world his place should be ready for him. His home needs to be decent, however humble and small, and his parents should be able to supply him with the necessities of life, however simply and economically.

The parents should have the ability, the determination and the wish to set an example in their own behaviour. It is easy to have a baby but not at all a simple matter to guide and mould the child into a worth-while person. Parents have to strive for this. It does not simply happen. Yet here is the joy and greatness of parenthood.

Religious teaching, practice and example should be a part of the home life. It is not enough to send children to outside places of worship such as church or Sunday school.

The mother

A home would not be a home in the full sense of the word without a mother. Undoubtedly she is the important person in a small child's life. There should be an intimate, secure, loving association between her and her child which gives them both happiness. Psychologists realize today that mother-love is probably the greatest influence in the formation of character and personality. So too, is the mother likely to be the greatest influence in the home. Even in this changing world in which we live today, the mother is the person who makes the home or who allows it to be simply a dormitory, some kind of hotel, a place with walls and a roof where her family come to sleep, eat and perform such things as washing and dressing themselves.

A child needs a father too

The father

A child can get something from his father which he can get nowhere else, from no one else. He needs both a mother and a father. In the early days of babyhood a father may feel out of place, awkward, unwanted. Yet if he cares to take an interest, to help and to take part in what has to be done, he may find a tremendous satisfaction and pleasure. As the child grows he comes to gain additional experience through his father, which he could never get in quite the same way from his mother. The father represents a strength and competence which added to the maternal love gives even greater security.

When a child is small it is difficult to think of him as adolescent, yet the time comes all too soon. The father who has established a friendship with his child from an early age will be more likely to keep his friendship and trust later, when it can mean so much to both.

Good habits

The child who is allowed to be slovenly in his habits is likely to be slovenly mentally and morally. Through forming certain good habits almost without knowing it, the character also is being formed. Good habits of cleanliness, sleeping, feeding and the things which go to make an orderly life, come to be performed without conscious attention being paid. How often is a very fine character found living in disorder and dirt? A child's natural love of approval can be used to encourage good habits.

Bad habits

Sometimes they appear because of laziness and a natural disinclination both on the child's and the parent's part. Or they may be the result of wrong handling making the child resort to something for comfort or in desperation. It will not help for the child to know that his bad habit is causing concern, distress, worry, and making him important. Whatever is done, however the bad habit is corrected, it must be handled in the right way, with a great deal of tact. If necessary, advice can be had from doctor or clinic.

Truthfulness and lying

A small child often 'tells stories' as we put it. This in fact is what he does. Until we are able to teach him where imagination ends and reality begins we should not be worried or angry if what he says is not strictly the truth. But we do have to lead him to this point. By the time he is school age he should know and understand that while pretending can be fun, it *is* only pretending and must not be confused with fact and truth.

There is no excuse for adults lying to a child about anything whatsoever. Even if the truth is unpleasant or unpalatable it must be given to the child. For example if the mother has to go out she must not say she will be back soon when in fact she will not be back for some time, in order to save herself a scene or the child some temporary unhappiness or upset. She should not promise anything she cannot fulfil.

Why should a child lie?

It could be to get himself out of some small trouble. He would not lie if he felt he could depend on his mother to understand. It could be because the truth would not gain approval for him and he was in need of approval. It might be that he was confused or under pressure of some kind. Perhaps the standards set for him were too high.

Whatever the reason, nothing will be gained by hasty scolding or horror and the thing to do is to get at the root of the matter, to find out *why* he feels he must lie.

Obedience and discipline

The child has to learn obedience in certain matters, to be disciplined so that he can learn self-discipline. He should know that his mother and father and other members of the family have their own rights and need consideration too. If he does not, he will have a rude awakening when he goes out into the world, for no one will want his company. He will be unpopular, even disliked.

If he is made to obey, amiably, in all important matters, from an early age, it should not prove too difficult. But insistence on obedience should be reserved for the important matters so that he does not suffer from continuous nagging over what is unimportant. It has been said that misbehaviour is when a child does not act in the way a particular adult thinks he should act, under particular circumstances, at a particular time. Very often this is the case and adults enforce obedience and discipline on a child without good reason. In the process of his training there will be numerous occasions when he has to be made to conform, when his behaviour is really bad. It is best to reserve orders for these occasions and not to waste them unnecessarily, for constant repressions may well result in the child refusing to obey when he really must—the whole thing becomes too much for him to follow.

The small child will be confused by being given too many reasons for obeying but between the ages of about three and five, much can be done by quiet talks and explanations. When a child can see clearly the reason for certain instructions he will more readily obey. When he is told to do something and asks why, it is reasonable to supply the answer and not to say 'Because I say so.'

The good meaning of the word discipline, used in connection with the training of children, is 'educating, instructing, teaching a mode of life'. There are bad meanings of the word too—and this brings us to punishment.

Punishment

This will not so often be necessary where a child is loved and loves in return. Where people are kind and co-operative towards the child he is more likely to be so to them. Yet it is unlikely that the child can be well trained without having at some time or other to be punished. Parents who want to avoid using punishment too often will have to be extra quick in order to exert firmness before rather than after the event. The child can often be prevented from committing the offence which would be punishable.

The reason for punishment should not be simply an outbreak of ill-temper on the part of the parent, the culmination of some annoyance. It should be given with a definite purpose, to accomplish something which cannot be accomplished in any other manner. If it does this and has no other serious effects then it has been right and good.

Punishment should not be threatened and then not administered if the threat has been ignored. Therefore it must be well thought out, reasonable, something fitting which meets the case. And the child must understand why he is punished.

Most parents would say that some punishment at some time is bound to be necessary. Necessary is the word to remember.

Punishment can sometimes be a bad thing. It can invite a repetition of the bad thing done. It can make a child lose confidence in himself so much that he will cease to try, feeling it useless. It should never be unfair, harsh, or vindictive.

Good temper and self-control

Again this is a matter of example. If the child sees endless instances of ill-temper and loss of control, he cannot be blamed if he behaves in the same way. Nor is he so likely to be cross if there are no real grounds for him being so.

Also ill-temper can become a habit and if, from his early days, he has learnt that displays of temper get him nowhere, he will not try to get his own way by temper.

Most children do lose their tempers and control at times. They must have their energies turned to something else.

Good manners

Life runs more smoothly when manners are good. Again this is simply a matter of example. The child cannot be expected to be polite if others are not polite to him, and there will be times when he will be forgetful and fall short of the standards set.

If he has been brought up to like people, to be unselfconscious, it will be easier for him. If his mother tells him to say this, that or the other, he may be unable to do it, yet if left to himself, may well make friendly overtures which are the beginning of what we call good manners.

A child should not be reminded or taught about his manners in the presence of others. This will only upset him and have far less effect than if he is taken away quietly somewhere and asked to listen to reasonable explanations.

Fairness, unselfishness and thoughtfulness

A child may be naturally fair but will have to learn to be unselfish and thoughtful. He is likely to have an inclination towards helpfulness, and by accepting help and showing appreciation, his mother can lead him further. As he gets older he can be told about consideration for others, can be asked to put himself in someone else's place. Once more example counts and much can be done through quiet talks and explanations.

Religion

Parents who have sincere views and deep beliefs will be able to convey them to the child. Whatever these may be, it is imperative to show tolerance of others and to teach the child that *there is good in every religion*.

What is meant by bringing up a child to be a good Christian

1. His home would be a true Christian home in every sense of the word.
2. Parents would be trying to lead a Christian life all day, every day, not only on Sundays.
3. The family would attend a place of worship together; and at home the child would learn to say Grace after meals, to say prayers, would sing hymns, have Bible stories told, pictures shown him and later would learn to read the Bible.
4. Full use would be made of the various festivals, to educate and to teach their meaning and to give him real happiness, e.g. the story of Christmas can be told or read many times. Children love repetition. Christ's birthday and how it has been celebrated and become a family festival because presents are exchanged in memory of Him, is something which can be understood by young children.

This time of year could be used to show practical Christianity too, e.g. the giving of ourselves and our time, not only material things.

Other festivals can be explained as the child becomes capable of understanding.

Supplement

Object of character training

The object of character training is to guide a child's mind into a knowledge of right and wrong and to train his will to overcome any impulses which he knows he should resist. Emotions play an essential part in training character. They must not be repressed yet cannot be allowed to dominate greatly. Love and curiosity, even fear and surprise, can be directed so that they are either good or bad. Such emotions as anger, hate and jealousy need to be treated carefully rather than suppressed through force, if they are not to warp the character.

An appeal can be made to the child's natural love of approval and each time that he behaves well makes it easier for him to repeat this good behaviour. In fact he forms a habit and does not consciously do one thing or the other.

Many studies have been made in recent years of the effects made on children by different ways of handling them. Certain points have emerged which are interesting. Here are a few of them:

1. It is unwise to offer a choice when in fact there is no choice, e.g. 'Don't you want to stay with Auntie while Mummy is out?' when in fact he *has* to stay with her. The direct command is best. 'You will be staying with Auntie while Mummy is out.'

   It is always best to be positive rather than negative, e.g. 'Please clear away your paints when you finish' rather than 'Don't leave your paints lying around' and 'Ride your bicycle carefully' is better than 'Don't have an accident.'

2. Shouting is of no avail. A calm, quiet, reasonable voice will have greater effect. When talking to the child at length, perhaps explaining something to him, the adult may make closer contact and a greater impression if she gets down to the child's level physically, or brings the child up to hers. To tower above hardly creates the right atmosphere of confidence.

3. Comparisons are more likely to make him dislike the person held up to him than to encourage him to copy him.

4. It is best not to make a child feel guilty or inferior but rather to explain in simple words *why* he should not have done the thing. It can even be said that others have made similar mistakes before they learnt better.

5. Any kind of dishonesty is foolish, even if it is a meaningless, mild, white lie intended to be comforting. (This does not mean that, as part of his training in good manners, the child need not learn that there are times when an unkind truth need not be blurted out.)

One way of planning how to form a child's character is to list the qualities it would be good for him to have, such as: kindness, thoughtfulness, gentleness, unselfishness, honesty, truthfulness, loyalty, understanding, tolerance, reliability, confidence, good temper, obedience, fairness, self-control.

Then his background and all the influences around him could be examined to see whether he is having every possible opportunity to develop these qualities.

Should he show signs of developing undesirable traits —bullying, quarrelling, lying, jealousy, hatred, or aggressiveness for example—then careful study of the possible reasons must be made and steps taken at once to improve matters, remembering that now more than ever is it important to treat him in a proper and calm manner, to remain on good terms with him, not to sever relationships in any way. This is the only way the child can be helped.

Appreciation of the child's view

How should he know right from wrong until he has been taught the difference? He is immature and inexperienced. If he commits what his mother thinks is a grievous sin, when in fact he had only made a mistake, then punishment would be wrong. In many ways children are completely at the mercy of the whims of adults. It is not easy for all of us to remember our own childhood or to have an instinctive understanding of the very young but we must try to develop the ability to see situations through the child's eyes. If this cannot be done it is unlikely that the child will receive entirely fair treatment.

CHAPTER 10

Childish illnesses

Nursing a sick child. Washing him in bed. Keeping him happy. Prevention. Hospital. Immunity. The common cold. Infectious diseases. Chicken pox. Mumps. Measles. German measles. Scarlet fever. Whooping cough. What to do. The premature baby. Twins. Adopted children. SUPPLEMENT. *Diseases. Smallpox. Diphtheria. Poliomyelitis. Giving medicine. Ear infection. Croup. Convulsions.*

A child may arouse suspicions that he is sickening for something by being slightly off-colour, off his food, fretful, tiresome or generally out of sorts. Or he may, in an alarming fashion, be playing and romping healthily one minute and quite suddenly become ill. The attack may have some simple explanation such as a digestive upset or it may be the beginning of one of the infectious diseases or of some serious illness.

Pain or violent symptoms of any sort would necessitate an urgent call for the doctor, of course. Whether the matter seems serious or not the child should be put to bed and any symptoms noted. Finding out if he has a temperature will be a useful guide as to his condition. How to do this is something which must be learnt in practical classes but is not at all difficult.

The normal temperature is 98.4, but this may go up and down slightly according to the time of day and what the child has been doing. A young child's temperature may rise after strenuous activity so it is best to take it after he has rested quietly for a time. It must be remembered too, that children up to the age of about five years old, may run quite high temperatures (104 and even higher) with slight illnesses such as colds, influenza, sore throats, and a high temperature may not mean a serious illness although a doctor should be called for a temperature of over 100 or if there is any cause for worry at all.

The sick child with a temperature should not be taken to the surgery. He should be in bed and a brief but intelligent report sent to the doctor who will then know how urgently his advice is needed. While waiting for him to come preparations can be made for his visit so that his valuable time will not be wasted. There should be a clean towel for him to wash his hands and the child can be prepared for him and kept cheerful. Food should not be offered or allowed in the meantime although drinks of any sort, if the child wants them, may be given.

Once the doctor has called, made his diagnosis and given his instructions, then it is up to the mother to see that these are carried out implicitly.

Nursing a sick child

If, during health, the child has been taught obedience and common sense in the things that matter, he is likely to be a good patient if handled properly. A spoilt child may be a very troublesome patient. Whatever methods have been used in his upbringing, this is not a time when he can be allowed to get his own way in any matter affecting his treatment, medicines or anything which might prevent or retard his return to good health. But he will need special care, sympathy and extra attention.

While carrying out the doctor's orders the child should be treated as normally as possible and encouraged to return to normality as soon as he can. Anxiety must not be shown or conveyed to him in any way. Optimism and cheerfulness about a quick recovery will help him as much as the doctor's medicine.

Washing him in bed

A sick child should be washed all over once a day and his face and hands sponged whenever necessary. How to give a blanket bath should be learnt in practical classes. His bed should be properly remade once daily and tidied and straightened as often as is necessary for his comfort. He will need frequent change of clothing and the room must be kept aired and fresh.

Keeping him happy

At first he may not want to play if he does not feel well enough, but he will appreciate and be helped by having company as much as possible. He will probably feel lively sooner than it is advisable for him to leap around in his bed, or in and out of it. When he has reached this stage then he should have quiet occupations which will appeal to him and keep him from being bored.

Although he should be kept happy, to aid and hasten his recovery if for no other reason, he should not have the whole household at his beck and call. The state of illness should not appear to him to be like one continuous party or birthday with everyone rushing round in circles at his bidding, but rather should it be something, slightly unlucky perhaps, which we all have to suffer at times, and need sympathy for, but which we all try to recover from as quickly as possible. The pleasures in store as complete recovery draws near should be greater than any he has had during the illness.

Prevention

Many childish ailments are avoidable. If the child is properly cared for in every way he is less likely to contract anything. If he does, he will recover more quickly than would an ill-cared-for child.

Germs enter the body through the nose and the mouth, through the digestive organs and into the blood-stream from a cut or scratch. Therefore the air that children breathe should be pure and fresh and they should not be near infection or taken into stuffy and crowded places. Everything which is eaten and drunk must be clean and fresh to avoid danger. Cuts and abrasions must be cleaned thoroughly to prevent poisoning.

Hospital

In the case of an emergency there might be no time to prepare the child in any way for leaving home and mother, and for facing this strange terrifying world. Therefore it might well be useful for a child to have some knowledge of hospitals according to his stage of understanding. Small children can play, with their dollies and Teddies, pretend-games about hospitals giving some sort of idea of the usual routine, uniform of doctors and nurses, regular visits of parents and the welcome home after recovery, with perhaps small rewards for good behaviour.

Older children enjoy this sort of play too, and for them it can be done more thoroughly, along with explanations about meals in bed, bed-pans, blanket baths, calling the nurse, being put to sleep, injections, masks.

If the word 'hospital' is not completely strange to the child there will be some slight advantage should he have to go there. If there is no emergency and there is time to prepare him more fully it would be well to do so according to his ability to understand. No lies should be told, the truth can be presented as far as is wise, in a matter-of-fact, sensible way, with emphasis on the enjoyable side of hospital life. There is in fact much that is fun in a children's ward. Stress can be laid on the importance of obedience towards the nurses and doctors and how much this will help recovery and hasten the time when he can come home. Doctors and nurses can be shown as kind people who can be talked to freely and who will be friends. Children who are old enough to think and to worry should be encouraged to talk over their fears and should be told whatever it is they wish to know in a tactful manner.

A few familiar and favourite toys, especially any usually taken to bed, should be packed. An older child might be thrilled to have a new pair of pyjamas or nightie, ready to wear after the hospital ones were no longer needed.

Any individual habits, foibles, might be confided to the nurse or sister who will be only too glad to know how she could help should he be unhappy. It would be a pity for him to have a sleepless night just because no one knew that he wanted a certain thing before settling to sleep, or if he was made miserable through no one understanding some word of his own, coined by himself or used only in the family.

A small envelope or parcel handed to him by his mother as she leaves him, to be opened when she has gone, often prevents a scene. The lending of a precious possession may give confidence—something of Mummy's which he can keep till Mummy comes tomorrow not only is a thrill but gives confidence in the certain return of Mummy.

It is really quite unnecessary for a period in hospital to cause a great upset if the mother will take some trouble. It can almost be made to look like some sort of treat! Hospital of course should *never* have been used as a threat. Nurses, doctors, dentists, hospitals and those in them should always have been treated with great respect and given due admiration so that the child has grown to think of them in a proper manner.

Immunity

Every child should be protected by the available inoculations. IN FACT IT IS ESSENTIAL TO ASK AT CLINIC OR WELFARE CENTRE WHEN AND HOW ADVANTAGE MAY BE TAKEN OF EVERYTHING OFFERED. It is important that accurate records are carefully preserved.

The common cold

A cold itself is not serious but sensible care has to be taken so that there are no complications. The child should be kept warm indoors for a slight cold and put to bed if it is at all bad. A persistent or lingering cough, a pain or a temperature, would necessitate the doctor being called in.

It is difficult to avoid colds entirely but the child who is leading a healthy life, living in well-ventilated rooms, being much in the open air, not being overheated either by too many clothes or too stuffy atmospheres, will be less likely to catch a cold.

Infectious diseases

From the time a germ enters the body till the first sign of the disease appears, is called the incubation period. Then there is the period of illness and finally the convalescence.

An infectious child should be isolated if at all possible during the time he is likely to spread germs. This varies with the different diseases. The following is a guide to some of the more common diseases.

Chicken pox
Incubation period: 11–21 days

Separate, raised pink spots appear on head, face and waist. They increase in number rapidly, forming nodules and blisters, then a crust which comes off without leaving a scar.
The child is considered infectious for fourteen days from the date of the appearance of the rash.

Mumps
Incubation period: 12–28 days

No rash but swelling, often painful, of the glands of the neck. Generally unwell.
Infectious for seven days after the swelling has subsided or fourteen days after it started.

Measles
Incubation period: 7–14 days

Rash appears on the face and behind the ears. There may have been the appearance of a bad cold before the rash. There is likely to be a cough, inflamed and watery eyes, rising temperature and the rash will be dark crimson spots. The child may return to normal life fourteen days after the rash appeared if he seems well. Doctor's orders must be carefully followed to prevent complications.

German measles

Incubation period: 12–21 days

Rash appears on the face. It is similar to measles but smaller and pinker. It is a much milder disease than measles and often the child does not feel at all ill. Infectious for seven days from the appearance of the rash.

Scarlet fever

Incubation period: 1–8 days

There may first be sore throat, vomiting, fever or a headache. Then rash appears in groin, sides of the chest, any of the warmer parts of the body. There will be small bright red dots close together, sometimes appearing to be a crimson flush. There will often be a high temperature, furred tongue, paleness round the mouth.

The child will probably resume normal life seven days after returning from the isolation hospital unless there are still discharges from spots, nose, ear or throat.

Whooping cough

Incubation period: 5–15 days

There is no rash. It starts with a dry cough and what seems to be simply an ordinary cold. Then comes the characteristic whoop followed by vomiting.

Quarantine period is difficult to state. It is usually until coughing has ceased. The doctor will say.

What to do

For all these and any other infectious diseases the doctor must be called in. The child should be isolated, other people with whom he has lately been in contact should be warned, and various steps considered. The room he is to be in should be kept as dust-free as possible. It should be well ventilated and with a minimum of hangings or materials which could harbour germs. It must be remembered that germs love heat, dust and stale air.

The child's laundry should be soaked in disinfectant before being washed. Utensils used by him should be kept apart from others, food left by him burnt. He should use paper handkerchiefs which can also be burnt. Whoever is looking after him should wear an overall while in his room and should wash hands before and after attending to him.

Disinfection after the illness is over may be adequately done by giving his room a thorough and extensive clean out, using disinfectants. The local public health department will give advice on this and it would be only sensible to consult them.

The premature baby

A baby weighing less than $5\frac{1}{2}$ lb. is generally regarded as premature. Certainly if he is as small as this at birth he requires extra care and probably special treatment. The doctor's or nurse's advice must be faithfully followed.

He must be kept very warm, and if he is too weak to

suck must be given his food in a manner which will mean little effort for him, perhaps by a medicine-dropper. If possible, breast-milk should be found for him and in the meantime his mother should be attempting to establish her own supply as it is of great importance for the premature baby to be fed on breast-milk.

He should be handled as little as possible and not removed from his heated bed in his warm room except for a very quick weighing twice a week. The scales should be placed near to his cot and he can be weighed rolled warmly in a blanket which afterwards can be weighed separately. While he is so small he need only be washed when it is necessary, and can just be wiped lying in his bed.

The first few days of his life are the most difficult as he is not really developed enough for life outside his mother's womb. He may stop breathing, or choke, but once the first few days have passed he is likely to survive and progress. While he is still less than $5\frac{1}{2}$ or 6 lb. in weight he is more susceptible to germs and infections but once he passes the 6-lb. mark he has proved, by doing this after starting at a disadvantage, that he has stamina.

So when he is 6 or 7 lb. in weight, and certainly by the time he is 8 lb., the parents must cease to consider him delicate or to treat him differently from other babies of his size. This is often a little difficult after having had to exercise such care while he was so small, but it is important to him not to be treated as delicate after he has survived his difficult beginning.

Twins

It will often be possible for a mother to feed both infants, but if not, she should see that they share equally. The ideal is for her to feed both at once (one at each breast) and with assistance this is possible. If there is not sufficient breast-milk, each should be given the complementary feed necessary after getting equal shares of the breast-milk.

The feeding of two babies at once, from the bottle, if a complementary feed is needed, presents practical difficulty unless the mother has a helper. It is quite likely that the two babies will also want the bottle at the same time and it may be that some compromise with our views on the subject of giving bottles is necessary. One method of giving two babies a bottle at the same time is for the mother to lie down with one baby on each side of her, propped in positions, facing her, which will enable her to hold their bottles at the same time. Or she can feed one baby entirely in the proper manner, in her arms, while the other is feeding with his bottle carefully propped, *beside* her. She may manage to arrange it so that one of her hands is free when needed, for the baby who is feeding by himself. If she does this, then the two babies must take it in turns to be fed either in her arms or alongside.

The thing to remember is that whatever methods are used each infant must have his turn, his fair share. If it is found too difficult to give both babies the complete bathing routine every night, then one can be skimped somewhat one night, and the other the next and so on. If the mother of twins can have help of course there will be no need to give either twin less attention than any other baby. A willing father can be a tremendous help here, for certainly the single-handed mother of twins will find it very hard to do all that she should, all the time, for two babies at once, especially as they are apt to want the same things at the same times.

Getting up wind may prove to be a worry if both are crying at once. One may be put to lie on his stomach for a few moments while the other is held—then they can change places.

Adopted children

When a married couple adopt a child the reason should be that they feel that they really and truly *want* one. It can be done through a registered society or the Local Authority. Both make excellent arrangements, are fully experienced in the matter of adoption and have sound judgement in the placing of particular children.

The child should be brought up in the knowledge that he is adopted, without stress being laid on it. He can be told how he was specially chosen because he was wanted and loved, that he gives great joy to his parents. The feeling that he should be grateful to them should be no stronger than that of any child to any parents.

Sometimes the question arises as to what need be said about the real parents. This is not likely to crop up seriously until he is old enough to understand about the strict rules of the adoption society, how they never bring real and adopting parents into contact. There is no way of finding out about real parents although the child may be comforted to know that parents are not allowed to give up children without the strongest of reasons.

It is worth remembering that the adopted child may have difficulties through the mere fact of being adopted. Therefore he, even more than any other, should be brought up to feel that he can confide in the parents who have adopted him. Special trouble should be taken to make him realize that he can bring his troubles and anything which puzzles or worries him, to his parents.

Supplement

Diseases

There should be no need to learn anything about smallpox but before the discovery of vaccination against it, it used to kill thousands of people in an epidemic and marked many others for life. It was indeed a dreaded disease. Today there is no excuse for anyone catching it.

Smallpox

Incubation period: 10–21 days

The rash is similar to that of chicken pox but will be on the arms, legs and face first and the patient looks very ill. There would have to be strict isolation and every precaution taken until he was pronounced well again, by the doctor.

VACCINATION AGAINST SMALLPOX SHOULD NEVER BE NEGLECTED

Every child should also have the diphtheria injections so that it too can be ruled out.

Diphtheria

Incubation period: 2–7 days

There is no rash but the throat will be inflamed, the skin very pale and there may be fever and sickness. The period of quarantine is usually three weeks but in the case of so serious an illness, it would of course be for the doctor to say.

Poliomyelitis

There is a vaccine against this too, which helps to build up the child's resistance to the paralytic form of the disease.

Incubation period: 7–14 days

There is no rash but the child would appear to have some sort of flu, perhaps stiffness of the neck, a headache or a general sick feeling. The fact that such symptoms *could* be the onset of a disease such as this emphasizes the need for calling in medical advice when there is any doubt about the child's condition.

Giving medicine

It should be given in a matter-of-fact way. Perhaps the child's attention can be diverted while the spoon is popped into his mouth. Tablets can be crushed into powder and mixed with fruit juice, fruit purée, sugar, honey, syrup or jam. It is best not to give them with whatever the child has in the usual way so that, if he objects to the powdered tablet it will not put him off the thing it has been mixed with, e.g. his mug of milk or orange juice, or in his cereal. Although it can be disguised in a flavouring he may like, it will be best to dissociate it from anything he has to have every day.

Eye ointment, drops or anything unpleasant can often be done during sleep. If not, the proceeding should be made as pleasant as possible. The older child may prefer to put the ointment on himself or even put the drops in himself and may be able to do so with the aid of a mirror, or with Mummy's hand simply to guide and steady him. The main thing is to avoid upsetting the child when he is ill and to be as tactful as possible, while at the same time carrying out the doctor's instructions accurately.

Medicines must never be given except when ordered by the doctor. Each illness a child has must be diagnosed separately and because some medicine ordered was effective for one illness it might not necessarily still be right for the next illness though to the mother the child's symptoms appear to be the same.

Ear infection

It is quite common for children to have mild infections of the ear. Whenever there is earache the doctor must be consulted so that he can prescribe the drugs necessary, especially if there is fever. While awaiting his arrival the pain can be relieved a little by applying heat (hot-water bottle or hot pad) and giving some Aspirin according to the age of the child. With prompt treatment with the present-day drugs it is unlikely that ear infections will develop into an abscess or mastoid.

Croup

This can be quite alarming for it may come on quite suddenly during the evening. The child, who has seemed to be quite well, or perhaps has only had a slight cold, wakens with a violent fit of hoarse, barking coughing which we call croupy. He may struggle and heave to get breath, and of course the doctor should be called.

Until he comes, the thing is to make the air round the child moist, or to take the child to where it can be made moist. A very hot water tap running soon makes steam, or a kettle or pan of boiling water. The child will soon be relieved if he is held close to the steam. When he seems better the mother should stay with him and keep the atmosphere round the cot moist.

Convulsions

It is more frightening than dangerous when a child has a convulsion. Usually it will be over quite quickly. He should be kept from harming himself. He may be biting his tongue and something should be placed between his jaws. He should be sponged and kept clean, comfortable and quiet until the doctor comes.

CHAPTER II

Play

Toys. Points in general about toys for small children. Baby, from two or three months old up to nine or ten months old. Respect for toys. Parents and play. Telling stories. Guiding a child's reading. Television. Music. Importance of duties. After about the age of six. How to feel at ease with children. Baby talk. Fantasy. Older children.

Toys

When a child is playing with toys, he is engaged in what, to him, is a pleasurable activity. He is amusing himself. But at the same time he is *learning* through play. Apart from keeping the child happy and occupied, toys and play are important. Nowadays manufacturers consult experts in child education so that they put the right toys on the market, and they are planned and made so that the child will be helped both physically and mentally. Yet this does not mean that he must necessarily play with them in the way intended. He should be free to use them as he wishes.

Points in general about toys for small children

1. There must be no sharp edges or anything which can

hurt, nor parts which could become detached and swallowed.
2. Toys will often be put in the mouth, so there must be nothing poisonous and colours must be fast. Painted toys may be poisonous.
3. They must be strong and well made, practically indestructible, weatherproof when possible, and easily cleaned.
4. They should be simply constructed so that they do not frustrate or make a child angry.
5. They should give scope for imagination.

Baby, from two or three months old up to nine or ten months old
At first while the small baby learns to focus his eyes he needs bright things to look at, then he will follow moving objects with his eyes and look from one to another. He will play with his own fingers, try to grasp anything put in his hand. At about six months old any pretty toy he can hold will appeal to him, such as a duck or animal. He likes noises too, and will love biting and chewing, so that rattles, rings, beads will be good. His toys will teach him a great deal, especially if he has a variety of different shapes, colours, sizes, weights and textures; and control of his hand and arm movements will increase.

At about ten months to about eighteen months old he may still like some of his first toys but will also get joy from more advanced ones. He will be able to hold and cuddle larger soft toys and will like such things as bricks, blocks, beakers, floating toys for his bath, and what are called fitting toys.

It is interesting to watch a baby with his first fitting toy. It may perhaps be one with coloured pegs which fit into a solid block. At first he may simply pull out all the pegs, throw them away and show no more interest for the time being. At this age a child will seldom play with one toy for long though he will return to the same toy frequently. Next time he may pull the pegs out, throw them away and chase after them. Eventually he may decide to return them to their places in the block. This takes some skill, control and concentration.

As soon as he can move about in some way, crawl, or walk, he will need toys that do more than encourage the use of his hands alone. Something he can roll along, and scramble after would be suitable and as soon as he is really toddling, push-and-pull toys will please him and have the added advantage of helping to improve his balance.

Anything which involves simple manipulative skill is good (provided it is not too difficult for him), things he can get in and out of will appeal, toys that open and shut, make noises, stretch, pop up and down, or in and out.

From about eighteen months to about three years old many of the toys mentioned will continue to be popular and useful. They will have taught him some fundamentals, developed him greatly as an individual and helped both his muscular and nervous system. But now he must advance further. He will become more inquisitive, more imitative, and imaginative. So his toys must allow him to be satisfied on these lines.

Bricks which interlock, screwing and unscrewing toys, posting boxes, water and sand and patty pans, will be popular. For the boy who wants to imitate Daddy, such things as harmless hammers, hammering pegs into holes, a wheel-barrow, a watering-can, small gardening tools— for the little girl, a teaset, from which innumerable cups of tea may be poured and handed round, a miniature carpet-sweeper, iron or pastry board.

Bricks may be made into a house one day, a train the next or almost anything the child happens to think of. If the materials are at hand the child will invent his own games to suit himself, his mood and his stage of progress.

From about three to five years old many of the old favourites will still be enjoyed, or perhaps more complicated editions of them. For example, a little girl will now begin to take care of a doll and will love to dress and undress her, bath, wash, comb her hair, put her to bed or in her pram. Boys (and girls too) may love to have their own gardens so that tools should be capable of work as well as play. Imagination will be even stronger now and any toys which suggest pretend-games are good, e.g. suits, uniforms, and outfits for dressing-up, model shops, garages, houses, etc.

Other suitable toys for this age would be clay for modelling, Plasticine, pencils and crayons, blackboards and chalk, paints and paint-books, alphabets, toy clocks or watches, mosaics, hoops, skipping-ropes, bats and balls, scooters, small tents, packs of cards, small handbags, pretty aprons.

Apart from learning and developing through play, psychologists believe that a child can be helped to solve his problems. As he grows older, more is expected of him, he has to behave correctly, he must stop being a baby. Although this sounds reasonable the child may have difficulties. Through his play he may be able to create situations similar to the ones he finds himself in, only this time he is master. This alters his angle of approach and may help him to conform in the way demanded in real life.

Respect for toys

Although it has been shown that toys are more than mere playthings, it seems necessary to point out that children should be taught to respect their toys. For although a child may love his toys, some more than others, they are very often treated with contempt by adults whose example he will follow.

It is quite common for an exasperated grown-up to shout 'Oh, go and play with your *toys*. You have plenty of them.' Or toys lying about may cause someone to kick them out of the way or swoop them up in ill-temper. Or the exhausted mother may say 'Put some of those *toys* away', the word 'toy' holding a sound of contempt. Toys are too often mentioned in a tone which belittles them, and sometimes toys which do not appeal are forced on a child. One adult may like to play golf, another tennis. Why then should we expect a child to like to play with a toy simply because Grandpa or Uncle Bill has spent a lot of money on it?

One child may be content for a whole morning laying out a farmyard, garage, or doll's house. This might bore another child who might prefer something he can handle, mould or hammer.

To ensure that toys are not treated as trivialities it is best for them to be kept in a place provided for them. Only a few should be available to the child at a time, and some might be kept as 'special', only to be handed out on request. Also, when a child stops playing with one toy he might be asked to return it to its place before being given another. At first, of course, he will need to be helped, but to take good care of his toys can soon become a matter of habit and routine if he is set a good example.

The four-year-old child should now be enjoying, or beginning to enjoy, play with others in his own age-group. He may still be aggressive, selfish and unsociable at times, but some time before the age of five he should be able to adjust himself to playing with his friends fairly well. As soon as he does, he will learn another very useful lesson in life. He will find that certain kinds of behaviour will make him popular and other kinds are likely to mean that he becomes ostracized.

We can help a child by providing him with, or sending him where there are materials and equipment which encourage group activities, e.g. slides, see-saws, certain types of swings, balls, several drums or 'musical instruments' placed near each other. We can start off games, demonstrate how to play certain things, but we should never force or make games too organized at this stage.

Generally speaking, *by the time a child is five years old* his life should have provided him with suitable play in sufficient space and freedom, to practise bodily control and at the same time allow for plenty of rest and relaxation. He should have had interesting and stimulating materials so that he could express himself and gain new experiences, yet he should always have gone along at his own speed, not having been urged into anything beyond his own ability. He should have learnt how to get along with other children, through play, how to make himself an acceptable member of a group. He should also have had some release of his emotions, through play.

Now, at school, he will have to learn some hard lessons. Games with rules may not at first appeal to him and he may find it boring to wait for his turn and too difficult to hide his bitter disappointment if he loses. But such games will form his character and he will soon learn the lessons they have to teach.

Also, from now onwards, constructional toys will be good for children who like them, and those games which help spelling, counting, reading, etc.

Parents and play

As well as providing the proper amenities for children's play by themselves and with other children, it must again be emphasized that parents can also do much by playing *with* their children at times.

The necessity for a 'Mothering hour' in infancy will be remembered and this is something that the growing child continues to need, although the form it takes will change. A mother should always have some time to spare for her children, apart from providing for their physical needs—to join in their little games, to play some part in their child's world.

Both boys and girls love to be with their *father* and to do things with him. At first the play between them might be more in the nature of romping and fun, growing into what can more properly be called games and finally becoming a tremendously enjoyable sharing of interests.

For the family to be together, to 'play' together is a most satisfying and valuable way of life to show the child.

Telling stories

Stories should be short and simple with plenty of action and drama. A young child likes illustrations and any adult can keep him happy with a story-book, allowing time for the pictures to be looked at and discussed.

The *telling* of stories is different and also has great value. Nowadays so many of a child's experiences are visual—television, comics, cinema—that if we can tell stories and hold interest in the telling, the child is being given yet another, different experience. It would seem a pity for the child to grow up, not really enjoying words alone, to

Sharing father's interests

feel that he must always have what are called visual aids.

Other advantages in telling rather than reading a story are:

1. The person telling it can 'act' a little, being more free without a book to hold. Children love this.
2. The language used can be adapted to suit the listeners.
3. Bits can be added or left out, according to the mood, to the child's responses or interests.
4. Somehow it brings the child and the adult closer.

One point must be watched. Telling a story seems to heighten its effect very often. It becomes more real, more personal and true. So that it is best to beware of anything too exciting which might result in bad dreams or too sentimental which might make a tender-hearted child cry.

Guiding a child's reading

When a child reaches the stage of being able to read for himself, parents can still take an interest and help him to find books which he will enjoy and yet which will add to his knowledge of life. Books need not be bought haphazardly. Guides and reviews can aid selection.

If the story telling and reading in his early days have been of a good standard it is not likely that he will be tempted to read only rubbish now, although he may be temporarily intrigued by comics or any poor literature he sees his friends reading. Parents who have shared reading interests with him, and given him many hours of pleasure with *their* choice of book, will find that their opinions are more likely to be respected. After all, their *likes* have been accepted previously, so surely their *dis*likes might now have a certain amount of influence. But they must not, of course, simply scorn to look at the comic or whatever the child is reading. They should ask to see it and then offer an opinion, anticipating that it will be at least listened to.

Television

Like many other things television can be harmful, if viewing is indiscriminate or excessive to the exclusion of other activities. But it also can have great value. As a form of recreation shared by the family it can be good, when the programmes are suitable. As an entertainment for the child who needs to relax for a short time, or to be kept occupied while Mother gets tea ready, it can be useful. Some programmes undoubtedly also have educational value. *But parents must decide, and stick to their decisions, what kinds of television entertainment, and how much*, their children are to have. For it is most influential in forming a child's attitude and beliefs, because of its visual realism.

Music

Babies love to be sung to sleep, sometimes a fretful baby will go off to sleep if soft music is played. The small child will enjoy singing, particularly 'action' songs, and dancing

She loves helping Mummy

or moving about to music. A percussion band will always be popular and parents need not be clever musicians to inspire their child with the love of music by having it in some form or other in their homes. It can be another interest shared, another part of the child's education begun.

Importance of duties

No one should grow up feeling that it is his right to be waited on hand and foot. Each individual should learn to look after himself and his possessions, to contribute suitably, according to his age and ability and the necessity for it, to family life.

First the child has to learn to wash, dress, clean his teeth and care for his person generally. Then he will be able to take on responsibility for his own belongings, his clothing, possibly making his own bed and keeping his room tidy. If the relationship between the parents and child is good he will *want* to help by going on errands, setting or clearing the table, fetching wood, etc. The carrying out of small family tasks is very good for the child in every way although he should not have burdens placed on him which he is unwilling to accept.

How children, from about three to six years old see their parents
A boy may be deeply devoted to his mother. He is likely to feel that there is no one in the world like her and it is common for him to declare that he will marry her when he grows up.

Yet by now he is beginning to understand that he will grow up to be a man and naturally this gives him a special interest in his father. Anyone observing the small boy during this period will find him copying his father to an extent which indeed should make any man see the necessity for setting a good example. What his father says, the words he uses, the tone of his voice, all his actions, his likes and dislikes, his treatment of others and his way of living—everything about him will be noted, consciously or unconsciously, by his son.

The little girl is likely to adore her father, but she too is at the stage of beginning to understand that she will grow up to be a woman. So she is likely to imitate her mother as much as possible, in the same way as the boy does his father.

After about the age of six

The child will continue to love his parents deeply but he now ceases to show his affection openly as often as he used to. He will become less dependent on his parents and increasingly interested in the views and ideas of other people he meets. Perhaps he may form an admiration for one of his teachers whose every word will become law. Or he may have some hero whose influence for the time being will appear greater than any.

All the same, the foundations have been well and truly laid by now. What he has learnt and absorbed from his parents in the early days has not been lost. It is all there and has made him what he is. His basic ideas of right and wrong have been so firmly ingrained that they cease to be his parents', but quite simply they are his own, part of him.

How to feel at ease with children

Some adults like children yet do not know how to make friends or get along well with them. All that is needed is common sense and a little understanding and willingness to try. Suppose some grown-up suddenly finds himself in charge of little ones for a few hours during an emergency. She, or he, may have a great deal of theoretical knowledge of children yet not feel at ease.

The small child may still be at the stage of coining words, may live in a world of fantasy and his opening remarks may be quite unintelligible. Yet if everything is accepted naturally, attentive interest shown in words, actions and any objects pointed out or brought, very soon the child will be ready to be friendly. He must be accepted as he is, on terms of equality, without condescension and his pride must never be insulted.

Baby talk

The child may speak in his own language and the grown-up may respond in the words proper for him to use. This will seem quite right and there should be ample understanding. Speech is, after all, our method of communicating with each other and we need not ask more of it for the purpose of making friends.

Fantasy

Any grown-up who is capable of taking part in a child's fantasy, and is invited to do so, may feel that the hand of friendship has indeed been extended. There need be no hesitation in taking part for the time comes naturally in all proper childhood when images and reality are separated and the child learns the difference between truth and lies. The opportunity for pointing this out to a child quite often comes at the end of some pretend-game and the grown-up who has participated in the pretence will be in a strong position of trust and friendship and well able to clarify the situation.

Older children

Conversation and friendship between older children and grown-ups also seems difficult for some people. The child may now be shy and unwilling. The best way is not to think of the child as a child particularly but simply to meet him as an individual and behave simply as one would to anyone else one meets for the first time. Conversation can be whatever comes to mind—probing for mutual interests perhaps, a discussion of sport, films, television, world affairs, holidays, absolutely anything—until there is a spark of interest, then the two can enthuse together.

A child may like to talk about his school life and friends. Such natural chatter will reveal much and help to show where points of interest lie.

Most important of all, the grown-up should be simply herself or himself, when approaching a child. There is no need for some special manner to be put on which it is imagined is suitable for children. Badinage is not usually popular unless it is in fact the normal way of behaving to all. Wheedling, coaxing, joking, over-friendly, over-familiar or silly attempts at conversation will not be appreciated. A clumsy waggish bearing is almost worst of all, for jokes and fun shared should arise spontaneously. Success will be assured if the pace of familiarity with each other flows easily, if no liberties are taken until friendship is firmly and certainly established.

CHAPTER 12

Speech and childish difficulties

Speech. Baby talk. Stuttering and stammering. Nail-biting. Fears. The fatherless child. The working mother. The handicapped child. How to give him the desired attitude. Answering questions.

Speech

This has not been given special attention yet a brief review should be of interest.

The meaningless sounds of early infancy, the small single words at about a year old (or many months later) gradually grow into something more intelligible. He has been communicating all the time in his own way and by the age of about two years he can often use two or three sentences. At three to four years old he should be able to converse with others reasonably well. Certain sounds present difficulty to different children. Some may find it hard to pronounce 'th' or 'r'. Others may not be able to discriminate between similar sounds such as 'k' and 't', or 'th' and 'f'.

As always, the way a baby is handled makes a difference. If his every want is anticipated he will have no need to try to ask for anything. If his mother is silent when she deals with him he has less to copy, or if she is too talkative, he has little chance. He needs the opportunity to talk yet should not be repeatedly urged to do so.

If a child's speech is not clear, or is oddly clumsy, by the time he is about four then a speech expert should be consulted, for whatever the reason it should be attended to before he has to hold his own with other children or he will find school life difficult and even cruel. Patience, skilled guidance and kind correction will be needed where there are genuine difficulties.

Defective hearing could be a cause but it is unlikely that this should not have been suspected sooner.

Baby talk

Where there are no genuine speech difficulties, baby talk is usually the fault of the parents. It may be that they cling to his babyhood as a time that they have loved and hate to see slipping by, it may be that the child himself does not want to grow up and continues the language he has always used which brings the attention he wants. It is up to the parents to speak sensibly, clearly and distinctly. The sounds the child hears should be the right ones for him to copy. The choice of words and phrases lies in the hands of those around him. And if baby talk fails to produce the response he wants then he will soon talk in the way which will produce it.

Certainly the little one experimenting in speech may coin words which seem lovable, amusing, clever or unique but it will be better for his sake to correct him. It can be done in a friendly joking way, without nagging or criticism, but he must understand that a correct way of talking now will please grown-ups more than any little tricks of speech he may produce, hoping to gain attention or raise a laugh.

Though 'family' words or phrases may be used, the child must be able to distinguish clearly between them and proper conversation, in public. At school he must have an adequate vocabulary and the ability to express himself as others do.

Stuttering and stammering

It is known that a child's emotional state has much to do with this and it usually occurs between the ages of two and three. It could be that the child simply tries to say more than he is yet able to. He may not know the words to convey his meaning. Very often the matter rights itself. The best thing is to appear to ignore it while trying to find the reason. If he has been tense or upset the cause should be removed. If he has been talked at too much, or urged to talk, then he should be left alone. He might be encouraged to *do* more and talk less for a time. When he does speak his parents should attend carefully and try to understand without his having to make too great an effort.

Another reason could be that he has been forced to use the right hand when he is really left-handed. The part of the brain which controls the hand is connected with the part which controls speech and it is thought that this can be a cause of stammering.

In any case no good will be done by direct correction or pressure. It is necessary to get at the root of the trouble and if the matter seems serious, or tends to become chronic, it is most important to get expert help.

Nail-biting

There is no direct cure for nail-biting itself. The cause has to be found. Coaxing, begging, nagging, punishing, bitter tastes on his nails—these are of no avail, for the child may be unaware that he is doing it. There must be something preventing him from being relaxed and at ease. The usual questions can be asked, e.g.: Is he being asked to do more than he is capable of? Is he being scolded too much? Is he unhappy in any way? Is he over-excited? Frightened? etc.

Almost all problems require similar treatment—the careful probing to get at the root of the matter. The same causes may produce different problems in different children and the only cure for all is to remove the cause.

Fears

Courage is a quality which is universally admired and grown-ups often show intolerance towards childish fears. A mother may not be pleased if her little boy appears to be timid or if her little girl does not like the dark. She may feel guilty or ashamed on their behalf, unwilling to admit that they are showing signs of not being brave.

Fear, after all, is simply an instinct to preserve oneself from possible harm. Certain instinctive movements are a natural defence against danger. It is then, not only a human failing but necessary for our survival. Nevertheless it is not possible to flinch from, to run away from, every threat of danger, and the child has to learn to keep his fears under control, and to take the line of denying fear will be useless.

When a child has fears, some concessions should be made to them until he is able to control them, e.g., he may have a night-light or his bedroom door open if he is afraid of the dark. Light or noises from the rest of the house will not keep him awake as much as his fears will. Or if he is afraid of dogs he need not be in close contact with them although his mother may make a point of going to talk to them. If he is afraid of water at the seaside he should be left to play in the sand or in small pools while others enjoy the sea and he looks on.

Psychologists have found that sometimes a child who is forced into something which terrified him may appear to react happily and satisfactorily. The mother may be pleased to think that by using a little firmness she has made him overcome a stupid fear, while all the time he has simply transferred his fear to other things. Outwardly tough, because he has no choice and does not dare voice his fears, he may still be nervous to a degree which can increase until he has great problems to face.

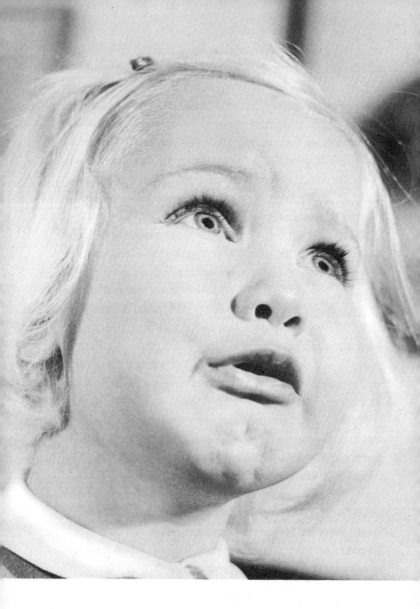

Fear should never be a matter for joking

The child will also often put himself into another person's place. If someone is injured, crippled, or dies, he (the little egoist) wonders if it could happen to him. His questions have to be answered to his satisfaction to stop his worrying. The use of strong imagination on top of unsatisfied curiosity can well produce great alarm.

An overheard conversation, possibly in hushed and awed tones about death, can fill him with terror. 'What is dying?' he may ask. He must be answered. Probably something like this would keep him from worrying: 'It's difficult to explain and no one really understands much, but we all have to die some day, usually when we are very old. You'll understand when you are grown up. There's no need to worry about it till then.' Some parents describe death as 'falling asleep'. There may be a danger here that the child will connect it with his own going to sleep and it might be better to talk of it as 'something like going to sleep' with the emphasis always on it being easier to understand when he is older. This prevents him from feeling immediate danger.

Fears should never be a matter for joking, or impatience, nor is it any good trying to urge him out of them. If he wants to talk he should be listened to with sympathy, comforted and reassured without scorn or disapproval.

The fatherless child

If the father is away from home or has died, it must be admitted that his absence is a great loss to the child. Yet the mother can do much to make up for it if she tries, and there is no reason why the child should not grow up in a perfectly happy and normal manner.

However hard it is, the mother should try to remain cheerful and sociable and avoid building her whole life and affection on the child. A boy especially will need the company of other boys and from the age of three onwards ought to have some contact with men. An uncle, grandfather, or friend may be available and it should surely be possible to find an understanding man somewhere who would spend a little time with the boy. Girls should not grow up either without knowing some sort of substitute or substitutes for their father.

His memory of course may be kept alive by photos and talking about him. Should he be only overseas for a few years then immense amount of time and trouble can be taken to keep close contact by letter. This can prove to be of great pleasure and value to all concerned.

The working mother

A mother may *have* to go to work to make a living and if she can make good and proper arrangements for her children this may be all right. She should try all the same, however hard it is, to give her children as much as possible towards the fulfilment of the needs (both mental and physical) talked about throughout this book. If the mother does not have to work but only wants to do so, for a little extra money, there is no question about it that her family should come first and what she can give is of more value than money or anything it can buy. Once they are at school it is different, provided she does not take on any work which would prevent her being at home whenever they are and caring for them and the home in the proper manner. A home in the true sense of the word, made by the mother and her presence in it, is worth more than all the extra luxuries her earnings could buy.

The handicapped child

Whatever the handicap, the parents will naturally be upset about it but they must try not to let the child see this. Many handicaps can be helped or overcome so that they need not spoil a child's life but this will not be possible if his attitude (which he will take from his parents) is wrong.

If the handicap is something like blindness, deafness, mental slowness or a serious physical one, then he will be under the care of experts and the parents should be guided by them for the child's good. Although such a child will need care, attention and help which his parents cannot be qualified to give him, he will still need them too, like any other child, and every effort has to be made to ensure that he gets the best guidance for his handicap without losing his parents.

The lesser handicaps, such as a slight physical deformity, cross-eyes, unsightly birthmark, and so on may well be cured at some time in his life if proper advice is taken and in any case should not be allowed to spoil the child's life.

A very slight handicap may ruin one person's happiness from his childhood onwards, while another seriously handicapped person may be completely unselfconscious and make a great success of life in spite of it. It is up to the person himself, and he will be helped if the parents have instilled the right attitude into him from an early age.

How to give him the desired attitude:

1. By giving him absolute love and security.
2. By treating him as naturally and normally as possible, making no difference in his upbringing except in so far as his disability might require it.
3. By making light of his misfortune and refusing pity. Yet showing sympathy through the efforts made to help and improve matters in every possible way.
4. By showing him how fortunate he is in other ways.
5. By praising his good points, searching for them, fostering and encouraging them, and giving him *something* to be proud of.
6. By making him feel useful, needed, and an essential part of the family.
7. By seeing that he mixes with others from an early age. The earlier he gets used to remarks and stares about his disability the easier it will be.
8. By taking him about, never hiding him or his disability or being ashamed of it.
9. By never worrying, fussing, criticizing or urging.

Answering questions

The sort of questions which only require patience or knowledge to answer are fairly easy to deal with. Parents can school themselves to patience, turn to other sources of information when necessary, such as an encyclopaedia or some more knowledgeable person.

But other queries, charged with emotion, about such mysteries as God, religions, birth, life, death, moral and social customs and our relationships with each other, are a more difficult matter. On the whole we may have no definite knowledge to offer in our answer, no facts, only an opinion.

When a child is curious about what a parent may consider an awkward subject, too often he is given a long and complicated reply, with much more information than he asked for. It is best to answer exactly what is asked, if he wants to know more he will ask more. It must be recognized too that if he is seeking information he is ready and old enough to be given it or he may feel that his parents have failed him or are inadequate. If he gets the feeling that he is prohibited from discussing certain topics he may become one of those who persistently ask aimless questions.

When the child is told what he wants to know, he himself builds up the whole picture in small digestible pieces. It would be much more difficult to explain things like birth, death, suddenly, when he reached some age decided on as being the right one. His knowledge should have grown with him.

INDEX

Figures after 's.' denote pages in Supplements

Accidents 84–8, 91–6, s.96–8
Adopted children 162–3
Affection 1, 122
Afterbirth s. 19
Aggression, Aggressive, -ness 173, s. 130–1, 148
Animal pets 78–9
Answering questions 191, 195, s. 82
Ante-natal care 3–4
Appetite 125, s. 115
Artificial respiration s. 98

Baby, clinging to mummy s. 68
  fretful 47, 178, s. 31, 32, 33, 66
  talk 77, 182, 186
Baby's room 13
  timetable 22, 23
Bath, -ing 8, 22, 23, 28–9, 73, 161, 170
  blanket bath 152
Bed 11–12, 120, 122, 152, 180, s. 82
Bedclothes 88, 105
Bedding 11–12, 13
Bed wetting 109–10, s. 112–13
Behave, Behaviour 22, 71, 72, 122, 135, 142, 154, 173, s. 111, 146
Birth 8, 20, s. 15, 18–19
Bites 95
Bone, -s 97–8, 100, s. 18, 114, 115
  broken 92, 97–8
Book 175–8
Bottle 41–6
  -fed baby 23, s. 34, 51
  -feeding 41–2; of twins 161
  -feeds 42, 44–6
Breakfast 60, 61, 72
Breast 22, 39
  -feeding 36–42, s. 19, 49–50
  -fed 48, 59, s. 34
  milk, *see* Milk
Breathing 39, s. 98

Bruises 92
Burns 85, 91

Calories s. 115
Carbohydrates 100
Cereals 48, 58, 59, 60, 61, 62, 126
Character training 134–46, s. 146–9
Chicken pox 157
Christianity 145, 195
  see Character Training
Circulation s. 98
Cleanliness 2, 5, 62, 128, s. 16, 34
Cleansing 43
Clinic 4, 8, 21, 127, 156
Clothes 10, 106, 156
  night 108
Clothing 106–8, 128, 152, 180
  inflammable 85, 107
  maternity 4
Cod liver oil 46, s. 114
Comforting 91, 94, s. 148
Commands s. 147
Common cold 151, 156, 157, s. 166
Companionship 47, 55, 128, s. 65
Confidence 129, s. 147
Confinement 20
Constipation 8, 100, s. 34
Conversation 47, 183, 185
Convulsions s. 97, 166
Cooking 102, s. 114
Cot 12, 118
Cough 156, 157, 158, s. 166
Crawling 54, 170, s. 68
Croup s. 166
Crying 23, 24, 27, 78, 93, 118, 122, 127–8, 129, s. 53, 69
Curiosity, s. 82
Cuts 86, 93, 153

197

Dentist 4, 102, 155
Diarrhoea s. 34
Diet 48, 99, 100–2, s. 82, 114–15
   additions 45–6, 48–9, 57, 61
   expectant mother's 6
Digestion s. 33, 115
Dinner 60, 61, 73
Diphtheria 156, s. 164
Discipline 141, 142
Diseases 156–8
Disinfection 159
Dislocation 93
Doctor 152, 155
   accidents 92–6, 105, s. 96–7
   bad habits 140
   bed wetting s. 112
   breast feeding 37, 40
   crying 122
   diarrhoea s. 35
   eye discharge 104
   feeds 44, 46
   illnesses 150–1, 156–9, s. 164–7
   playing at 154
   pregnancy 4, 6, 8, 21
   refusing milk 127
   vomiting s. 34
Dress, -ing 77, 107
Dried milk, *see* Milk
Drink, -s 48, 57, 59, 60, 72, 91, s. 113
   expectant mother's 7
   nursing mother's 38
Duties 180

Ear, *see* Hearing
   infection s. 166
   object in 95
Education 142, 180
Eggs 101, s. 114
Embryo s. 17
Emetic s. 97
Emotions 70–1, 174, 187, s. 130, 146
   emotional associations 136
Eyes 27, 28, 104–5, 169, s. 165
Exercise, baby's 27, 106, 123–4, 128
   mother's 2, 5, 6, 38, s. 16
Expectant mother 2–8

Fairness 144
Falls 86, 92
Fantasy 182, 183

Father 17, 22, 26, 84, 135, 138, 139,
   162, 175, 181, 191, s. 82, 83, 131
Fatherless child 191
Fats 63, 100, s. 115
Fears 123, 155, 188–91
Feeding, artificial 41–2, 48
   breast, *see* Breast-feeding
   complementary 40–1, 161
   self-demand 24, 26
Feeding, over-feeding 126, s. 52
   under-feeding s. 32, 53
Feeds 58–9
   making up 44–5
   quantities s. 49–53
   stopping late feed 47, 48
   temperature 45
   see Milk
Feet 105
Fire 91
Fires 85
First Aid 91, 95, s. 98
Floors 86
Fontanelle s. 31
Foods 57–63, 73–4, 99–102, 125–6
Footwear, *see* Shoes
Fractures s. 97–8, *see* Bones, broken
Fresh air 2, 5, 14, 106, 118, 123–4,
   128, 153, s. 16
Fruit 61, 62, 101, s. 114
Fruit juice 59, 62, 72, 73, 74, *see*
   Orange juice

Games 63, 73, 174, 175, s. 82
Garments 10–11
German measles 158
Germs 45, 62, 153, 156, 159, 160, s. 34
Growth 99–100

Habits 2, 27, 76, 108, 109, 120,
   139–40, 144, 155, s. 16, 69, 110,
   146
Handicapped child 193–4
Health 1, 105, s. 16
Hearing 27, 106, 186
Hiccups s. 32
High chair 106
Holiday 90
   travelling 63–4
Home 1, 2, 135, 136, 145, 193, s. 113
Hospital 154–5

Hot water bottles 85, s. 32
Hygiene 44, s. 15

Illnesses 150–63, s. 163–7
Imagination 140, 169, 171, 191, s. 82–3
Imitate, Imitation 171, 181, s. 81
Immunity 156
Incubation period 156–8, s. 163–4
Independence 76, 129, s. 133
Indigestion s. 32
Infectious diseases 156–9
Inflammable 85
 non- 107
Inoculations 156
Isolation 156, s. 163

Jealousy 71, 128, s. 112, 131–3, 146, 148

Labour s. 18
Lavatory 72, 77, s. 111
Laxative 8, 94
Learning 168
Left-handed 187
Lies, Lying 130, 140–1, 183, s. 148
Love 37, 42, 71, 118, 128, 136, 137, 142, 194, s. 133

Manners 143
Marriage 1
Measles 157
Medicine s. 165
Milk 38, 42, 44, 45, 46, 49, 62, 63, 73, 101, 126–7, s. 18, 51–2, 114
 breast 30, 37, 40, 41, 160, 161, s. 49–51
 dried 42, 45, s. 52
Minerals 100
Mother 137, 138
 –'s attention 24, see Spoiling
 expectant 2–8
 nursing 38
 working 192
Mothering hour 23, 47, 175
Movements 27, 54, 76, 106
Mumps 157
Music 173, 178–80, s. 32

Nail biting 188
Napkins 9, 29–30, 108, 118, s. 34, 112
Navel, s. 18, 31
Nose 28, 29, 39, 77
 -bleed 95–6
 object in 95
Nursery rhymes 64, 73
Nursing 151–2

Obedience 141–2, 151, 154
Orange juice 46
Out-of-doors 89–90, 107, 123
Over-feeding, see Feeding

Parents, see Mother and Father
Perambulators, see Pram
Pets 78–9, 88
Placenta s. 17, 19
Play, -ing 47, 63, 72, 73, 76, 77, 129, 152, 154, 168–75, s. 65, 66, 80
Play-pen 14, 47, 55, 88, 89, s. 69
Poisons 86, 94, 169, s. 96–7
Poliomyelitis 156, s. 164
Posture 105–6
Pot 108, 109, s. 80–1, 111
Pram 12–13, 22, 23, 47, 89, 118, s. 32, 69
Pregnancy 3, 4, 7
Premature baby 159–160, s. 18
Problems, see Behaviour, Emotions, Tantrums
Proteins 99, 101
Punishment 142–3, s. 149
Push-chair 105, 124

Quarrelling, see Behaviour, Emotions, Tantrums
Questions 191, 195

Rash 157, 158, s. 163, 164
Reading 177–8
Religion 137, 145, 193
Rest, child's 24, 64, 73, 74, 105, 121, 128, 174
 mother's 5, s. 16
Rickets s. 115
Right and wrong 135, 182, s. 146, 148–9
Roughage 100

Routine, child's 26, 173, s. 80
    mother's 5, s. 80, 133
Run, Running 70, 106

Safe, Safety 55, 64, 72, 84–9
    in cars 64, 90
    on holiday 90
    road safety 78
Scalds 91
Scarlet fever 158
School 77, 121, 174, 193, s. 113
Security 24, 37, 42, 73, 128, 139, 194
Self-control 143
Separation from mother 77, 130
Shock 91, 92, 93, 95
Shoes 2, 5, 105, 106, 107
    footwear 86
    rubber boots 107
    sandals 108
Sick child 151
Sitting up 54
Sleep 5, 22, 23, 47, 55, 73, 74, 105, 116–24, 128, 155, s. 16–17, 32, 69, 113
Smallpox 156, s. 163
Snacks 63, 74
Sociable 76–7
Social life 6
Speech 27, 71, 76, 182, 185–8
    difficulties 186–7
Spoiling, Spoilt 26, 47, 152, s. 33, 65–9, 80
Sprains 92
Squint 104
Staircase 86
Stammering 187
Sterilizing 43–4, 46, 62
Stories 121, 145, 175–7
Stuttering 187
Suck, -ing 38–40, 43, 49, 160, s. 18, 65
Suffocation 88
Sugar 44, 62, 63, s. 34, 51, 52
Sun 114, 124–5, s. 16
Sunburn 95
Sympathy 28, 152, 153, 194
Swallowing objects 86, 94
Sweets 62, 74, 125

Tantrums 71, 72, 129, s. 80

Tea 60, 61, 73
Teats 41, 42, 43, 44
Teeth, 2, 4, 37, 72, 73, 100–4, 180, s. 18, 114, 115
Television 128, 178
Temper 128, 143–4
Temperature 150–1, 156, 158
Tetanus 156
Thyroid gland 100
Time-table 22, 23, 26
Toddler 71, 72, 89, 123–4, s. 79, 80, 132–3
Toilet requirements 8–9
    requisites 28
    training 46, 108–10, s. 110–13
Topping and tailing 29
Toys 47, 55, 85, 86, 88, 120, 121, 155, 168–74, s. 69, 80, 81
Training 78, 109, 142, s. 111
Travelling 63–4
Truth 154, 183, *see* Lie
Tuberculosis 156
Twins 161–2

Umbilical cord s. 17, 19, 31
Under-feeding, *see* Feeding
Unwillingness to take milk 126–7
Unselfishness 144

Vaccination 156, s. 163, 164
Ventilation, *see* Windows
Vitamins 41, 45–6, 99, 124, s. 114
Vomiting 93, 94, 158, s. 33–4, 52

Walk, -ing, 55, 70, 73, 90, 170, s. 68, 69, 82
Washing 28
    himself 72, 76, 77, 180
    in bed 152
Water 41, 46, 48, 59, 62, 101, s. 52
Weaning 41, 47, 48–9, 57–64, s. 64–5
Weight 30, 57, 159, 160, s. 33, 49–51, 52, 53
Whooping cough 156, 158
Wind 22, 23, 24, 40, 118, 162
Windows 2, 5, 28, 86, 87, 123
Words 55, 70
Wounds 93–4